D0464557

LONG-DISTANCE FLYING

The Pilot's Library General Editor, Robert B. Parke

FLYING AIRPLANES: THE FIRST HUNDRED HOURS by Peter Garrison
FLY ON INSTRUMENTS by George C. Larson
LONG-DISTANCE FLYING by Peter Garrison
PILOT'S GUIDE TO PREVENTIVE AIRCRAFT MAINTENANCE
by J. Mac McClellan

Coming Soon:

WEATHER REPORT by Robert B. Parke
PILOT'S AVIONICS HANDBOOK by George C. Larson and Robert Denny

Other Doubleday Aviation Handbooks:

ANYONE CAN FLY by Jules Bergman
PILOT'S NIGHT FLYING HANDBOOK by Len Buckwalter
THE PILOT'S GUIDE TO FLIGHT EMERGENCY PROCEDURES
by Alan Bramson and Nellie Birch

LONG-DISTANCE FLYING

BY PETER GARRISON

With an Introduction by Robert B. Parke, General Editor

A PILOT'S LIBRARY BOOK

1981 Doubleday & Company, Inc., Garden City

Library of Congress Cataloging in Publication Data

Garrison, Peter.
Long-distance flying.

(The pilot's library series)
Includes index
1. Cross-country flying. I. Title. II. Series:
Pilot's library series.
TL711.L7G37 629.132′5217

ISBN: 0-385-14595-0
Library of Congress Catalog Card Number 80-2049

PRINTED IN THE UNITED STATES OF AMERICA
FIRST EDITION

CONTENTS

FOREWORD

Most pilots cherish ambitions of expanding their abilities and experiences. They wish, in the cliché of the industry, to "trade upward" to bigger and more expensive airplanes, or to "climb the rating ladder" to more difficult, arcane, and prestigious capabilities or endorsements. They wish also to go higher, faster, and farther than they have gone before, and they experience a childlike thrill the first time they climb past 20,000 feet, exceed 200 miles an hour, or fly across the continent. To have done all these things in a jet airliner, obviously, means nothing to them. They wish to have done them *themselves*, with their own hands upon the controls. Absolute speed, height, and distance mean little (though there is a coterie of those who pique themselves upon having flown faster than sound as passengers on SSTs), compared with one's own personal records.

I recognize this impulse in other pilots because I see it in myself. I am not a professional pilot and have not had the stimulus of higher salaries and better commands to propel me from rating to rating. I have kept climbing because in some peculiar way I am left uneasy by the thought that when I have only a single-engine rating, there are others who have a multi; when I have land, they have sea; when I have fixed-wing, they have rotorcraft. I suppose I imagine them to experience some satisfaction which I am denied; and so I go out and get the multi-engine rating, the seaplane rating, the rotorcraft rating, and so on. It is not so much that I am tormented or even tickled by envy, as that I am curious and don't want to miss anything.

I think most other pilots share that feeling. They do not gladly stop expanding their horizons. They want to go farther from home, in bigger, faster airplanes, flying night and day, in good weather and bad. To expand their horizons, they have no better means than self-instruction and experience.

No one disputes the fact that instrument training, for example, provides only a scratch recipe for safe bad-weather flying. The seasoning of experience is the final, and the most important, ingredient in the making of an instrument pilot. The same is true of private pilots; the private license, it is often said, is only a license to learn.

Each pilot has his own frontiers. Flights which seem everyday to one pilot might be memorable milestones to another; it depends at what point on the road of his experience each pilot finds himself at the moment. Far down that road from the first short trips of the new private pilot are the long hops of the businessman crossing half the continent in a single flight, or the pilot who decides to take himself, rather than be taken by an airline, to Europe, Alaska, or Hawaii. Long flights, ones that explore the limits of the capabilities of private airplanes, are what this book is about. They are different, not only in scale but also in the kinds of problems they present and the strains they put on the pilot, from short trips in slower airplanes taken close to home. Training airplanes usually cannot fly more than three hours before landing with a small fuel reserve; larger types, on the other hand, may have ranges of six or seven hours and more than a thousand nautical miles (nm). They may tax the endurance of the pilot, and expose him to vicissitudes of terrain and weather and hazards of fatigue which he is much less likely to encounter in a short-ranged Tomahawk or a 152.

The longest flights of all—transoceanic flights—involve challenges new even to most experienced pilots. The easiest ocean flights—those from Florida to Yucatan, for instance, or from the Mexican mainland to the peninsula of Baja California—introduce pilots to sensations of insecurity and anxiety which they never encounter over land. Fear, in an airplane, is dangerous in itself, and the pilot's own emotions are one of the hazards he must prepare for. Longer trips, such as that across the North Atlantic via Greenland and Iceland, can still be made by some aircraft without extra fuel supplies, but they encounter official requirements for survival and communications equipment unfamiliar to most pilots. Still longer flights, such as Atlantic crossings from Gander directly to Shannon or the Azores, or from California to Hawaii, require supplementary fuel systems and entail navigational difficulties without parallel in land flying. At the same time, ocean flights raise the challenge of fatigue and deteriorating competence just as long overland flights do.

In the twenties and thirties, when the pioneering long-distance flights were made, they sorely taxed the capacities of both airplanes and the pilots. Airplanes have made a lot of progress since then, but pilots have probably degenerated. Today long-range flights are accessible to anyone who cares to try them, not only to a small corps of dedicated heroes and publicity-hounds.

I will define a long-distance flight as one of more than four hours' duration; this is a very generous definition, four-hour flights being quite common and certainly not regarded by most pilots as particularly extraordinary. To the extent that sheer time spent aloft, with the fatigue and the change in weather conditions that it can bring about, is significant, four hours probably represents a fair threshold of "long" flights. On the other hand, in some ways it is distance, not time, that defines a long flight, and in this case I would put the lower boundary at 600 or 700 nm. At the extreme, flights of 3,000 or 4,000 nm, and fifteen or twenty hours' duration, are within the reach of modern light aircraft, with minimal modifications, and of their pilots, similarly modified.

Long flights can involve, besides material hazards, certain psychological pressures and pitfalls. The longer and more ambitious one makes one's flights, the more the completion of the flight begins to appear as an end in itself. The idea of numerous en route stops becomes repugnant; one does not want to spoil a fast time, or descend from a high altitude, or miss an intended dinner or appointment for some insignificant reason; and so one is tempted to dismiss progressively greater and greater reasons as insignificant. Part of gaining experience in flying consists of losing our fear of ephemeral dangers such as night or rain; but part of the mortal danger of flying results from our inability to distinguish between illusory and real hazards when we have not yet experienced them.

What makes a four or five hour flight differ from a two or three hour one, and 700 nm a different kind of distance from 300 nm? A combination of several elements. One is weather. If the weather is good, a long flight may be not much different from two short ones. In bad weather the difficulties multiply, not only because the weather can change significantly over a period of a few hours, but because fatigue takes a greater toll of us during instrument and marginal VFR (visual flight rules) flying, especially during the final phase of the flight, when we are the most tired and the greatest demands on our skills may be made. Not only weather, but also terrain can change; flying from the Great Plains to the Rocky Mountains, for example, one encounters gradually changing conditions, of which density altitude is the principal one. Air temperatures are lower, and freezing levels closer to the ground. At the same time, the terrain suddenly becomes hazardous as one crosses the eastern boundary of the mountains. Ceilings and visibilities which were not threatening over the plains become so over the mountains.

A pilot might fly for several hours from an inland area to a coastal zone, and after hundreds of miles of CAVU weather (clear or mostly clear) find fog or stratus obscuring his destination. Not that this could not happen on a short flight; but the weather briefing for a short flight will be more up to date, whereas on a long flight, a pilot who does not periodically update his weather information may find a surprise awaiting him when he arrives. He will be tired and low on fuel, and the destination will just have gone under.

Time itself is a factor, in the form of the motion and position of the sun. You may take off in the morning eastbound and arrive after dark, having crossed three time zones and devoured most of a day. Flying westbound, you may stare for hours into a burning afternoon sun. The time of day affects the weather; evening will bring a lowering of temperatures and an approximation of the temperature and dew point, and may trigger the beginning of rainfall or the formation of fog. On the other hand, it often happens that the first morning light causes fog to form in a place that has been clear all night. Afternoon brings daily thunderstorms in many areas; flying eastward, you shorten the day, and with it the gap which separates dawn from the development of thunderstorm conditions.

Foreseeable developments, like the growth of afternoon cumulus or the formation of fog in coastal zones after midnight or in early morning, merely require you to make intelligent plans in advance and to take off at such a time as to avoid arriving at exactly the wrong moment. In the case of morning fog, you should plan your arrival for late enough that the sun will have "burned off" the fog or lifted it to form a stratus deck with good visibilities underneath—commonly around ten in the morning. In the case of thunderstorms, which frequently persist into the night, you have to take care to be at your destination, or at the location where the storms will develop, early enough in the day. As the day wears on a line of towering cumulus gets denser; the passages through it for VFR flying are fewer and fewer, and soon even IFR (instrument flight rules) penetration gets risky, especially without radar.

Thinking ahead in time can get confusing. Weather reports are given in GMT (Greenwich Mean Time), also called Zulu (Z) time. American pilots may experience a certain difficulty in thinking of diurnal weather variations in Kansas in terms of a clock located in London. You know that the cumulus will start to build on the inversion layer at about ten in the morning, that by noon one or two anvil heads will have started to form, and that by three or four you may be forced into significant deviations from your planned route. You also know that it won't get dark till around eight, and so you'll be able to pick your way through the storms for quite a while. If you're going to be westbound, the prognosis is better; because you'll be flying toward the light, gaps in the clouds will be more apparent, and the day will be longer-lasting. But that kind of thinking comes from experience with the diurnal variations in weather in the summer in the Midwest. When you look at the weather report, you have to convert your instinctive grasp of the weather pattern to the irrelevant numbers of the GMT twenty-four-hour clock. The only time GMT is really convenient is when your destination is London.

In addition to the opportunities for error which are produced by the alien time-frame in which weather reports are given, there is the inherent uncertainty of the reports themselves. In general, the shorter the time between your getting the forecast and your using it, the less the likelihood of error, though there are plenty of ways for even a short-term forecast to go wrong. When you take off on a flight of six or seven hours with a destination in the area of a stalled front or a slow-moving warm front, the meteorological balance is delicate, and the possibility of changes in the weather forecast while you are en route is great. At such times, not only is it important to keep abreast of fresh weather forecasts; it is also necessary to note the changes in successive forecasts, to estimate the confidence level of the forecasts, and to be particularly careful in monitoring not only the destination airport but also the alternates. Very long flights are likely to bring you near the end of your fuel supply, and the availability of the alternate may become critical.

Long flights do not tax an airplane mechanically any more than short ones do; less, in fact, since they represent extended periods of stable operation, without bursts of high power or changes in temperature. The engine does not get tired from running for five, ten, or thirty hours straight, though the pilot may. Supplies of consumables become a consideration, however. Obviously, there is the fuel supply—the first concern in any flight. There is also oil, which may become a consideration when the engine is old or burns or pumps a lot of oil, and oxygen, which is often the limiting factor on long high-altitude flights. Sometimes oil and oxygen run out neck and neck. In general, turbocharged engines pump an increasing amount of oil out the breather at higher and higher altitudes. Pilots are often not aware of the loss of oil, since they may not be accustomed to making long flights at high altitudes, and other factors, such as nosewheel strut extension or ramp slope, affect dipstick readings and introduce a certain randomness into the periodic addition of quarts of oil. Most airframes, too, have the breather extending sufficiently far from the fuselage or nacelle to prevent the oil which is going overboard from leaving its trail—which can be immense—on the airframe; so one can easily be unaware of a high rate of oil consumption at high altitude.

Discomfort and fatigue play a role in long flights that is quite out of proportion to their influence on short ones. Discomfort seems to increase exponentially with time, especially at the moderately high altitudes where supplemental oxygen is not yet necessary. Soreness, back pain, and headache may appear. Noise is thought to be fatiguing, though it has been hard to demonstrate experimentally that it is; and I suspect that to the extent that body fatigue occurs on a long flight, it is exacerbated by vibration. Few traveling salesmen drive Volkswagens. Many people are sensitive to lack of food or drink; dehydration and low blood sugar are likely to appear on very long flights. There may be some deterioration of visual acuity from long-term shortage of oxygen.

In addition to physical afflictions of various degrees of severity, night flights in bad weather and overwater flights introduce the additional elements of anxiety and fear. These are sensations which some pilots have never experienced; those who have not are surprised to discover, when eventually they do, how their judgment, instinct, sense of time, and even their ability to interpret simple instrument data are degraded by anxiety. The French speak of "mental viscosity"—a vivid expression which evokes the weird thickening and slowing down of the thought processes. A flier may get a fright on any flight, however long or short; but overwater flights in particular, with their lack of navigational aids and sometimes of any kind of contact with human voices, can bring about a sense of panicky loneliness and doom which takes the happy-go-lucky novice by surprise.

Fear produces several ominous states of mind, one of which is the obsessive desire to get onto the ground. Over water, this is obviously an idle wish, but it brings about a distortion in one's sense of time—always a part of airborne anxiety attacks—which lends a false urgency to every act. Even though there is nowhere to go and all the time in the world in which to act, one feels impelled to make snap judgments and to take some kind of action immediately. At such times the best thing to do is usually nothing at all.

Attacks of fear are a specific hazard of ocean flying, until one has made sufficient crossings to have become casual about them. Not that the first is the worst, and crossings necessarily get better thereafter; they all involve instrinsically frightening situations, and the pilot may make a couple of long overwater flights without a second thought, and then find himself sinking into a panic on the third or fourth one because of some unexpected problem.

Fear may be resisted in several ways. For the beginner, optimism, deliberate avoidance of thoughts about danger in the abstract (What if the engine quits? What if . . . ?), keeping busy with navigating and bookkeeping, studying approach plates, or reading books are the usual prescriptions. For the veteran nothing is new; he has seen all sides of most hazards and knows their actual magnitude. In between is a stage (through which instrument pilots also must pass on their way to becoming veterans) where situations arise which it is impossible not to take seriously, but whose true import the pilot cannot, for lack of experience, estimate. For instance, it inevitably happens to every pilot that he finds himself wandering around in cloud one day without radar, knowing that there are thunderstorms about. ATC (Air Traffic Control) isn't being helpful. Other airplanes are all around asking for steers around weather, ATC is too busy to comply, and, to top it off, the ride is smooth with only occasional light rain. What to do? Is this the situation they were talking about when they hammered into your head never to fly into embedded thunderstorms? If so, what are all these other pilots doing around here?

Experience bridges the gap between the acquired ideas of the new pilot and the instincts of the veteran. It is a gap which cannot otherwise be bridged; countless words will never take the place of a few minutes in the purplish heart of a storm cell, or in the middle of a boundless ocean.

In international flying, there are the complications of regulations, which vary from region to region; fresh difficulties in interpreting weather reports coded in an unfamiliar jargon; and the difficulty of access to general information about weather conditions, availability of services, conditions at remote fields, fees and charges, and so on. The fees themselves can be surprising, and bills have a habit of continuing to arrive long after you've returned home. On some ocean routes, flight planning may be influenced as much by landing fees and fuel prices as by considerations of distance or weather.

Some of the experiences of long flights can be quite novel; others are only exaggerations of what one can encounter on any flight. In flying, however, a matter of degree is not always trivial; it can grade over into a qualitative difference which in turn separates a successful flight from a near escape, or worse. *Almost* no icing, a fuel system that *nearly* worked, are not good enough; to the extent that any flight demands excellence, long flights demand something closer to perfection.

This book is about flying high-performance general aviation airplanes to the limits. It covers two categories of long-distance flying: operations within the United States, of the kind that pilots encounter regularly; and over-ocean operations, and those in foreign countries, which pilots encounter more rarely, but which present more urgent and unfamiliar problems. To the extent that editorial policies of magazines and the putative readership of books have permitted it, I have always tried, in my writing on aviation subjects, not to confine the field of my attention to the experience of novice pilots. Even in writing about basic concepts, like pilotage or engine operation, I feel that one ought not to purge the subject of its complexities merely for fear that they will make it incomprehensible to a novice. My own learning has often consisted of a first stage in which I read materials that were over my head, and later, from reflection and further reading, came to understand ideas that had first been opaque. In reading much aviation literature—and instructional literature in other areas as well—I have been struck by the amount of false information that creeps into it as a result of attempts to remain within the understanding of the least sophisticated reader. In the area of engine operation, for example, rules of thumb proliferate, both in articles and in the materials coming from the manufacturers, which not only deprive the airplane operator of access to the full capabilities of his engine, but even lead him to inferences about it that are simply untrue. In the cause of not confusing the beginner, they mislead the veteran.

In some chapters in this book, therefore, notably that on efficient flying, I have gone into greater technical detail than some might think necessary, or welcome. I have done this for the sake of those who are willing to follow the ramifications of the subject as far as possible, and at the risk of repulsing others who would like to have everything reduced, for their convenience, to two or three rhyming couplets or four-letter acronyms. Better, I think, to supply complete information than to supply a little and leave the reader to invent the rest. Much of the discussion on efficient operation is admittedly of an academic nature, for even pilots who understand the subject thoroughly end up operating their airplanes by a few simple rules. But the rules themselves are insufficient without an understanding of what lies behind them, and of the limits of their applicability. They are incomplete without their foundations.

It is particularly important, I think, to give flight efficiency a thorough treatment, because it has become an issue of ostensible importance in discussions of fuel shortages and of ways to apportion limited fuel supplies. General aviation, encumbered with the popular perception of it as a luxury of the very rich, depends, in support of its own survival, upon its fuel efficiency being at least the equal of that of automobiles. The mechanical efficiency of automobiles is gradually improving, however, while there is less likelihood that that of the general aviation fleet will improve significantly in the short term. Only through education of pilots to efficient techniques of operation can much improvement be hoped for here. I have consequently thought it important to give this topic a thorough, if in parts somewhat technical, development. Even if they do not choose to operate in the regimes of maximum efficiency (which have many drawbacks in terms of speed, comfort, and potential engine trouble), pilots should understand the boundaries of those regimes, the laws that they obey, and the gains and losses that departing from them involves.

I have also dwelt at length upon the subject of area navigation and upon some of the uses of certain programmable computers for route planning. These subjects seem to me to be important for two reasons. One is that various forms of area navigation equipment are filtering down into the general aviation instrument panel, and that an understanding of the concepts underlying area nav is going to become as important to all pilots in the next ten years as an understanding of the use of omni has been in the last ten. We will all eventually begin to think in terms of latitude and longitude and of direct routings selected independently of the Victor airways structure.

Since much of the work done by area navigation systems is computational, and since there is no reason not to think that the applications of microcircuitry to avionics will continue to proliferate as rapidly, at least, as they have done in the last few years, I think that it is well for pilots to become acquainted with both the capabilities and the limitations of computers. Already much of the newest and most expensive equipment requires fairly extensive programming; with programming, these sophisticated devices take over many of the navigational tasks which used to belong to the pilot. Rather than plot his course on a chart, or even note the VORs that picket his track in the margin of his note pad, the pilot punches lat/long coordinates into the "on-board computer." Soon all the tasks of clocks, fuel totalizers, power computers, nav radios, and annunciator panels will be handled by this on-board computer; and when this happens, it will behoove the pilot of the airplane to have some understanding of how the computer works, and what exactly it is doing.

I have included a section on using a Texas Instruments TI59 for flight planning both because it is quite handy for certain kinds of tasks, especially in plotting very long trips, and because it gives some insight into the mechanics of area-nav computers and some "hands-on" experience with thinking in terms of lat/long rather than VOR identifiers.

There is one aspect of this subject, however, about which I feel uneasy. What distinguishes the computer-nav attitude to flying from the VOR attitude, and the VOR attitude, in turn, from the pilotage/dead reckoning/NDB (Nondirectional Beacon) attitude, is that each represents a higher level of abstraction than the preceding one. The primordial kind of flying is the kind you do when you take off from your home field without ever looking at a chart, and fly to another field fifty or a hundred miles away to have lunch or visit with a friend. You never look at a map, and you don't have to check the weather, any more than you would when going out for a walk or a bicycle ride.

The next stage—the one most general aviation pilots are in—is the one in which you plan a flight on a symbolic representation of the world—the chart—and you do your navigating in advance. The navigating that you do in the air is only a verification of what you have already done on the ground. When you arrive at a mountain or a lake, it merely serves to confirm what the chart already told you would be there and what you planned for, in terms of altitude and routing, before you took off.

The next stage is one in which no surface detail is even needed. Instrument pilots get in the habit of using instrument charts even for VFR flying, because they are compact and easily managed, and they contain all the necessary information about VOR frequencies, airway bearings, and leg lengths. Safe altitudes are prescribed on the chart, and it is unnecessary to know the locations of mountains, or where there are rivers, lakes, or railroads; it is only necessary to know frequencies, bearings, minimum altitudes. The route is represented not by a pencil line on a relief map, such as the student pilot learns to draw, but by a list of three-letter identifiers or airway designators. The flight no longer concerns the pilot as a voyage over the earth, but only as the execution of a planned track through a matrix, the spinning of a line from point to point in space and time.

When flight planning is done on a computer, and turbocharged or turbine-powered aircraft fly at altitudes above the tops of the highest mountains in the country, the pilot need barely concern himself with any kind of chart at all, let alone with the features of the countryside. He needs lat/long for the origin and destination, and some frequencies en route. The trip has become totally abstract.

We still look out the windows, of course. But occasionally I regret the loss of the intimate feeling of the ground that one had at first, picking one's way from landmark to landmark, racing the cars on the roads. There are so many good reasons to fly high, that we forget to fly low; and in coming to use the ultimate simplifications of flight provided by the electronic computer and area-navigation system, we risk losing touch with what first made flying so beautiful—moving along above the ground, and seeing everything around us. The whole fun of it was the relationship to the ground; we were free, but we enjoyed our freedom because we were still in sight of our prison.

The kind of flying this book is about a paradoxical blend of the personally satisfying and the alienating. It is deeply satisfying to arrive in Ireland after a long flight from Newfoundland or Iceland; it renews your sense of wonderment, your awe, it makes you feel younger, makes you forget your worries. But to fly many hours at a stretch, making thousand-mile leaps, is also to skip everything that lies in between. Not much lies between Gander and Shannon; but between Amarillo and Los Angeles there is a good deal, and if you fall too much in love with your ability to go from one to the other in four and a half hours nonstop, you may forget that stopping is sometimes better than flying. I hope that in writing about the demands, the techniques, and the rewards of long-distance flying I will not forget, or tempt my readers to forget, that there is a great deal to be said for short-distance flying too.

INTRODUCTION

It's tempting when the weather is great and the airplane is working beautifully for general aviation pilots to dream momentarily of flying on, magic carpet fashion, to some remote land or a favorite foreign city or even a different world. Most of us, however, are quickly returned to earth by the need to engage in honest toil or by the realization that the idyllic moment will soon pass and a more realistic condition of horrid weather and headwinds will quickly overtake us.

Peter Garrison is one who seldom allows tedious reality to interfere with his savoring an appealing fantasy. He also has a rude curiosity about remote lands and a good many favorite foreign cities. It is likely he would be easily induced to look over a different world, particularly if he could fly there. A decade ago when the idea of staying aloft long enough to reach any of these destinations occurred to him, there were few airplanes that had the range or performance that he required and none in the price category that he had in mind. The solution for Garrison was to design and build an airplane that could cover several thousand miles nonstop and that would cost a fraction of what a comparable production airplane might cost.

It mattered little to him that his engineering background up to that time barely qualified him to change a light bulb or that he had never built anything more complex than a tree house. He presumed that most or all of the necessary information was in books and books were in libraries and there were plenty of libraries in southern California, where he was living. There were also in southern California a fair number of people who were building homebuilt airplanes as well as one or two highly skilled designers of kit airplanes.

Garrison was well into acquainting himself with the dark arts of aerodynamics and propulsion when I was let in on his intentions. At the time I was editor of a magazine for which he was writing with great frequency, and he chanced to mention to me one day that he was thinking about building an airplane and flying it around the world, and he asked how I might regard the story possibilities of such an adventure. I thought he meant to fly around the world nonstop but it turned out that he was planning to make the journey with many stops of indefinite duration and to spend several years doing it. I also assumed he was going to build or modify one of the kit airplanes that were on the market or to severely revamp a production airplane to meet his range and performance needs. But no, he insisted that he was well along with the design of his own airplane and that he was going to start construction in the backyard of his home in Tarzana, the following week. I assured him that if indeed he did any or all of the things he planned to do the story possibilities were excellent.

Shortly thereafter I was in California and while visiting Garrison I was invited to see the engineering drawings of the airplane. Feeling a bit like I was being shown the perpetual motion machine of a mad scientist, I followed Garrison into a back room where on a makeshift drafting table supported by a couple of sawhorses there were indeed some authentic-looking drawings of what appeared to be a rather short-coupled, low-wing, single-engine airplane. It had massive tip tanks, a smallish wing, and retractable gear. The cockpit was set up for two seats, and at the time the tail was conventional. Years later it was modified to a T-tail that became a trend-setter for general aviation production airplanes.

Over the next few years, *Flying* magazine reported on the construction, first flight, and subsequent long-distance flights of *Melmoth,* for that was its name. Less thoroughly reported was the fact that perhaps in part as a result of Garrison's successful flights each way across both the Atlantic and the Pacific, a good many more pilots than ever before were tackling major overwater flights and equally inhospitable land-mass flights using long-distance flying techniques developed by Garrison and others. It was as though the reliability of powerplants and the sophistication of avionics combined to encourage many pilots to want to tax their skills beyond the every day omni-to-omni flying. Beside the satisfaction of having flown across an ocean and the practicality of being able to use their airplanes for transportation at the destination, there was the undeniable sense of accomplishment at having taken a risk by stretching their range, endurance, and skills to the limit.

While overseas flights by private pilots are not yet commonplace, a little informed reckoning indicates that as many as a thousand single-engine general aviation airplanes a year span successfully one of the Atlantic routes or a major leg of the Pacific. At least that many more reciprocating-engine twins and turbine-powered business airplanes routinely fly to Europe, South America, Africa, and beyond. Increasingly, wherever one goes in the world there are U.S.-registered N-numbered airplanes to be seen lined up at international airports. And like as not one can see the haughty G-IIs and Falcons standing next to Bonanzas and Centurions.

True, a good many of the single-engine flights are made by a hardy band of professionals whose job it is to deliver U.S.-built general aviation airplanes to every corner of the world. It is the frequency of their flights that has made it easier for ordinary pilots to find suitable additional fuel tanks for almost every type of airplane and to find experienced mechanics to install them. This cadre of regular delivery pilots also fulfills the need of first-timers for details of the proposed trip and guidance on dealing with protocol. A few can also provide first-hand information on survival gear and how to handle potential emergencies.

It isn't surprising, with all the transoceanic traffic, that there is an occasional misadventure. The cause is often not what one might suppose, the sudden catastrophic engine failure that is ever present in the pilot's mind the instant he loses sight of land. Nor is it some unforeseen and inexplicable weather phenomenon, if any reasonable amount of planning has been done. More often, according to those who have made it back to land or those who have been plucked from the briny, and even from radio transmissions of those less fortunate, it is some auxiliary device or accessory that fails and does the flight in. The last couple of such events I heard about are perhaps typical at least in some respects.

The first involved an almost new and very expensive business jet that was returning from Bermuda and was headed for Atlanta. At the worst possible moment it became clear that a fuel valve was stuck and the awful possibility loomed of having to ditch as a result of fuel starvation while there were still several thousand pounds of fuel on board. By going to the most economical power setting and aiming for the nearest airport, the flight limped in and landed safely to the considerable relief of the crew as well as of the owner, who was presumably relaxing in the cabin.

The second case was a little out of the ordinary in several respects. It involved a single-engine airplane on a delivery flight in midwinter across the North Atlantic. On the trip to the overwater takeoff airport there had been a few glitches that might have made a more superstitious pilot nervous, but repairs had presumably been made and the pilot was a veteran with over a hundred successful crossings. Several hours out the airplane was cruising along at 9,000 feet when the engine stopped cold. The pilot carried a small life raft and a foam-insulated survival suit as well as other emergency gear. Attesting to his nimbleness and of course his high motivation is the fact that he was able to squeeze into the suit during the few minutes he had while letting down. He also managed a frantic Mayday to the Canadian rescue forces and was able to give them a reasonably accurate position. The actual ditching went well enough and two and a half hours later a rescue plane hove into view and dropped a more serviceable life raft and some flares. Nine hours later a weather ship picked him up and weeks later deposited him safely on land. Although he has temporarily given up delivering airplanes overseas, he continues to have confidence in the practicality of general aviation airplanes successfully making long-distance overwater flights and in the benefits of having an airplane available for travel about Europe and points east.

He, like others of our acquaintance, confirms that the most important virtue for general aviation travel outside the United States is patience. Abruptness and haste seem to arouse suspicion and generate resistance. Casualness bordering on lassitude is more in keeping with the attitude that is accepted in most parts of the world. Often a willingness to gossip will promote harmony and cooperation.

One memorable example flashes to mind. The trip did not involve a particularly long-distance flight but it is instructive in how to deal with airport personnel. The city was Florence, Italy, and the airport had a reputation of being bothersome in releasing general aviation flights. I had completed the international flight plan and was bound for London. The airplane was a twin Comanche with tip tanks and I was trying to get a weather briefing. I listened to the English-speaking pilot ahead of me get a thorough weather roundup in rapid Italian, which he apparently didn't speak or understand, and as he left I prepared to get much the same. I opened the conversation with the briefer, however, by admiring the Fifteenth Air Force plaque on the wall behind him. He replied in English flavored with an overpowering southern accent. It turned out that he had picked up his second language from the citizens of Alabama, Mississippi, and Texas who were stationed nearby during World War II. He appeared to delight in practicing his skill, and only after discussing conditions in Montgomery, Biloxi, Dallas, and other cultural centers of the south did we get to the weather. The briefing was thorough and, as it turned out, correct, and as it turned out the entire exchange had only taken an hour.

There are those who want no part of that kind of time-wasting and return to the states railing about the indifference of the foreigners to general aviation. Others, who can enjoy the curious practices of foreign officialdom and who contemplate the horrors of foreign airline travel—endless waiting, thorough searches, shuffling on and off buses, callous personnel—will take flying their own airplane any time.

But it's futile to belabor the arguments about the practicality of flying a small and vulnerable general aviation airplane for many hours across a hostile surface in order to enjoy the benefits of that airplane at the destination. The rationale for doing so is much less concrete and much more obsessive. In some remote respect there is often a compulsion to share with others who have gone before the sense of exposure to elements that could dash you down in spite of your best efforts. There is a yearning to experience challenge that is unique to pilots, when they have drained their mind in planning, testing, and considering every contingency, when they must know whether their skills are sharp enough and their resourcefulness deep enough, and there is no way of knowing the answer unless they go. And when after months or years the airplane is in takeoff position and the runup is over and there is nothing ahead but advancing the throttles and wondering how it will end—only then is it possible to appreciate, without necessarily understanding, what is happening. The moment to stop is gone and the airplane is rolling. There may be six, seven, ten hours of flying ahead and several weather systems to go through but for now the things to be done are the procedures and practices that have been rehearsed and that, regardless of the outcome of this flight, will make them better, more professional pilots than they would otherwise be.

When you are merely considering the prospect of a long-distance flight, the need to know more about your airplane than you can get from the pilot's handbook is apparent. In Garrison's case he kept meticulous records of fuel flow and experimented with unusual power settings. Although on most of his long-distance flights he used conventional settings, he knew that if the need should arise he had the information he would need for adjusting his power to achieve a specific result.

If there is a further lesson to be learned from Garrison's flights it is, Know thy airplane. While you will never understand your airplane as thoroughly as the homebuilder knows his, it behooves you, if you are considering flying an ocean, to know more about your transportation than just how to check the oil. For openers, follow your mechanic when he does inspections. Draw your system diagrams from memory. Shut down your engine (over a suitable emergency landing place) in the air and restart. And do as many things as you can think of in the bright light of a VFR day so that you can reduce the anxiety when something unforeseen occurs in lousy weather at night.

After following the evolution of *Melmoth* and watching Garrison's successful flights, I find myself wondering when long-distance flying in ordinary general aviation airplanes will be commonplace. The technology is largely here now. NASA is working on powerplants that will be more efficient than our present engines by a factor of two to three. They will be lighter and possibly use different fuels. Burt and Dick Rutan and others are experimenting with more advanced airplane designs. Navigation by satellite is within reach. Advanced autopilots that will automatically fly a thousand miles with enormous accuracy are a reality.

If production airplanes of reasonable cost, both singles and twins, could handle 2,000 nm at cruise speeds of up to 200 knots, think where you could go and what you could do. Everyone could then do what Garrison can do now.

ROBERT B. PARKE
General Editor

LONG-DISTANCE FLYING

1

THE CONQUEST OF DISTANCE

Airplanes were one of the great novelties of the early decades of our century. Their powers of levitation were not unprecedented—balloons had carried passengers aloft, and even for considerable distances, since the eighteenth century—but their speed and maneuverability set them apart from lighter-than-air devices. The aspect of their performance that exercised the greatest hold upon the popular imagination was range. Speed, height, duration of flight, payload carried were all significant, and pilots and designers strove to surpass one another in setting new records in each category; but it was long-distance flying that hit the newspapers, that catapulted some of its practitioners to fame and wealth, and that actually changed the world. The long-distance fliers of the twenties and thirties opened the air routes that now knit the world together; they brought the continents within reach of each other, and obliterated the boundaries of time that separated them.

It is worth reflecting, in recalling the early days of distance flying, that before the invention of wireless telegraphy the most rapid means of travel across bodies of water was the steamship, and that even along the most technologically advanced and heavily traveled routes, like the North Atlantic, the voyage might take a week or two. Wireless had had the effect of speeding the flow of information from place to place, but not the flow of materials or people. The expeditions which left Europe in the fifteenth and sixteenth centuries to explore the Western Hemisphere sailed away into utter darkness, so far as their families, friends, sponsors, and compatriots were concerned. They traveled for years. When they returned, most people had forgotten that they had ever left. The armies of empires, from the time of Rome and before, could at least send back runners, who might travel for weeks to inform the senate of a victory or a defeat; but by sea one could not send runners. When settlers came to the New World, the people they left behind might learn no news of them from one year to the next. That America was dubbed "the New World" was not remarkable; it may as well have been another planet, or another stellar system. A kind of principle of relativity, its basic unit the speed of a ship or of a man on horseback rather than the speed of light, ruled those separate worlds; and as we cannot know any events less than ten years old on a star ten light years away, so the world could not, before the development of electromagnetic communication, have had any fresh news of distant places. All news was old. Distances were distances in time as well as in space; and that temporal separation was the really significant aspect of spatial remoteness. It made events on other continents irrelevant and unreal, like events in television serials. Even the news arriving, up to date, by wireless from Europe in the late nineteenth century was of mainly academic interest to Americans, so thoroughly were they insulated, by space and travel time, from events abroad. The world at the turn of the century was still in many ways what it had been for thousands of years: a cluster of floating islands, remote from one another, barely fixed in one another's consciousness, bound by fragile and precarious ties; and a world in which one could still believe in wonders.

The airplane changed all that. It was eventually to bring the continents virtually within sight of one another; but even at the outset of intercontinental air travel, when each new record was won at the risk of life and limb and with a prodigious application of energy and fortitude, one could foresee that it would only be a matter of time before the oceans were entirely subjugated. Europe was moving toward America, Australia toward England, Brazil toward France. By a kind of aeronautical plate tectonics, the primordial process of disintegration, in which the original continent, Pangaia, had broken up into pieces which floated away to the places where we find them today, was reversed, and the dispersed parts were reunited. This global upheaval, though invisible and soundless, was apparent to people everywhere; without reflecting, they could understand that there was something very strange and significant in the fact, reported by the morning paper, that an airplane had just flown nonstop from New York to Paris, or to Berlin, or to Moscow. It was almost as though an airplane had flown to ancient Greece, or to the land of the dead.

It was World War I that brought into being airplanes capable of intercontinental travel. They were only barely capable of it, at that—cumbersome, slow, inefficient, underpowered. Their normal condition of flight—slow, unstable, uncomfortable—was one which we today would consider intolerable. All the sciences that go into the making of an airplane were relatively young; structures were shaky and uncertain, engines no better.

The first of the great postwar flights was that of a Vickers Vimy from England to Australia. The Australian government had offered a prize of 10,000 pounds—about $50,000—for the first flight from England to Australia in under thirty days. The distance is 8,600 miles by a great circle route (which runs, incidentally, over the southern tip of Sweden, north of Moscow, and then across the center of China). Not surprisingly, the aviators chose the longer route down the axis of the British Empire, through Egypt, India, and Singapore. The great difficulty of the early long-distance flights was the lack of suitable airfields and service facilities; hence the rather generous allowance of thirty days to complete a flight that should not have required more than a hundred hours in the air even for the cumbersome Vimy. A flying field, in those days and along those routes, was not an aerodrome; it was simply a field that had, by decree or by custom, been used by airplanes. It might be full of tree stumps, or submerged by rain, or patrolled by a bull. The two pilot brothers, Keith and Ross Smith, and their two mechanics, who worked on the engines during stops and slept en route among the spare parts in the aft cockpit, encountered surprisingly little trouble on the trip, damaging their airplane only once, not seriously, and never once suffering an engine failure. They made it in under twenty-eight days, in the late fall of 1919.

A Vimy was also the first airplane to cross the Atlantic nonstop. A prize had been offered before the war by the London *Daily Mail* for the first nonstop crossing in an airplane; at the time the idea was pure pie in the sky. After the war, however, it was obviously quite practical, and a whole group of airmen set out to do it. Several airplanes were specially built for the flight, some of them with ingenious features like jettisonable landing gear and a fuselage turtle deck which detached to become a lifeboat. The only practical way to make the trip was from west to east, given the prevailing winds, though one contestant, the Shorts *Shamrock,* planned to take off westward from Ireland; it crashed, however, before getting to Ireland at all.

The several entrants in the *Daily Mail* sweepstakes gathered in St. John's, Newfoundland, in the spring of 1919 to await fair weather and a chance to start. In the meantime the U.S. Navy mounted its own transatlantic flight with four Curtiss flying boats and a string of twenty-five destroyers to mark the way. When the airplanes proved unable to take off with their full load of fuel, each got a fourth engine added to the three already installed. Such were the uncertainties of early flight, and the drastic measures taken to resolve them.

One of the Navy airplanes, the *NC-4,* completed the trip, but made a stop in the Azores that disqualified it from the *Daily Mail* competition. When news reached Newfoundland that the American airplane was at the Azores, the British team of Harry Hawker and Kenneth Mackenzie-Grieve decided to take off in spite of the weather so that the first crossing could fall to the English. They flew for part of a day and all of a night, but their engine developed a cooling problem and eventually they had to ditch in the ocean, having located, after a long search, a freighter that could pick them up. The freighter had no radio, and England had given them up for lost when the ship finally reached port. Hawker and Mackenzie-Grieve received a welcome, such was the general relief at their survival, that surpassed even that which John Alcock and Arthur Whitten Brown, the men who successfully made the crossing and won the prize, would receive a few days later.

Alcock and Brown; no one has heard of them today, though their nondescript statues grace a roadside at London's Heathrow Airport. They took off on June 14, 1919, in the afternoon. They flew all night in icy weather in their open-cockpit airplane, in and out of clouds, climbing and descending in search of clear air; after sixteen hours they arrived squarely in the middle of the Irish coast, at Clifden. They had ample fuel left and could have continued to London, but low clouds hung over the Irish hills, and the prize was won; so they circled round and landed on what appeared to be a smooth green field, but turned out to be a bog. The airplane ignominiously upended itself; the two men climbed out weary, half-deaf, but unhurt. People came up to ask whether they had heard any news of the transatlantic flight that was said to be under way.

For an airplane to have crossed the Atlantic was remarkable; dirigibles did so with ease, however, and at a somewhat—though not much—better speed than steamships. Shortly after Alcock and Brown's flight, the British airship *R-34* made the crossing both ways in a total of 183 hours, without incident, prompting Brown to write, a year later, "It is very evident that the future of transatlantic travel belongs to the airship." This statement, so entirely incorrect as it would transpire, was motivated, no doubt, by a sense of the effort it had cost him and Alcock (who had by then been killed in an airplane crash) to make the crossing, while a dirigible could do it so effort-

Developed as a long-range bomber by the British during World War I, the Vickers Vimy was the first airplane to cross the Atlantic nonstop. Courtesy National Air and Space Museum, Smithsonian Institution.

lessly. Another long-distance flier, years later, was to comment, "These people who talk about how small the world is must not have flown around it lately." There was more effort to long airplane flights than appeared.

For several years after 1919 the record books remained undisturbed. Then in 1926 the French succumbed to a frenzy for long-distance flying. Between June and October of that year, four successive records were set by military pilots in Breguet 19s, flying southeastward from Le Bourget, first to Greece, then to Iraq, then to Russia, then to Iran, the last flight—from Paris to Jask, Iran—covering 3,352 miles.

The Breguet 19 was used by several French crews to set distance records. The designer gave his name to the "Breguet Range Equation" (Chapter 6). Courtesy National Air and Space Museum, Smithsonian Institution.

About six months later that distance record fell, this time to an American: the most famous pilot of all, and the most famous flight. Charles Lindbergh flew from New York to Paris on May 20–21, 1927.

Lindbergh was flying for a prize, like the Atlantic pilots of 1919; this time it was the Orteig Prize, for a nonstop flight between New York and Paris. With preparations taking place under the nose of the New York press, and several well-financed and well known pilots participating, there was a great deal of public interest in the contest. Reporters stopped at nothing in their efforts to obtain, or concoct, a story. It was a reporter who attached to Lindbergh the nickname ''the Flying Fool,'' because he planned to make the flight alone and in a small airplane; in fact, no one could have planned his flight less foolishly than Lindbergh did. Though he was the son of a former congressman, Lindbergh was neither rich nor influential; he raised the financing for his venture, and got his airplane built, by luck and determination. The flight itself—over thirty-three hours in the air—was a considerable ordeal; loneliness and fatigue in mid-ocean brought Lindbergh strange apprehensions and hallucinations, as they did to all pilots who made such flights.

Lindbergh's flight surpassed the former record distance by only 250 miles. But there was something special in his having made the flight alone, and in its having been over the Atlantic rather than the comparatively hospitable deserts of Mesopotamia; his youth, good breeding, shyness, and looks all contributed something. But it was the newspapers that made Lindbergh into one of the most famous men of his generation.

That Lindbergh had literally overnight achieved fame and, if not wealth, at least the certainty that he would never again be poor, made others want to do the same. With that fervor for repeating a good thing until it is worn out that we see in music and films today, newspapers, pilots, and manufacturers again and again, during the thirties, placed their hopes in this or that long-distance flight. There was never a second Lindbergh, so far as the public was concerned. But two weeks after Lindbergh's flight Clarence Chamberlin flew a Bellanca, with a nonpilot passenger, 3,910 miles from New York to Eisleben, Germany. A little more than a year later two Italians flew from Rome to the northern coast of Brazil—nearly 4,500 miles. The next year two Frenchmen flew 4,900 miles from Paris to China, and a year after that two Americans flew from New York to Istanbul. In 1933 two Frenchmen flew from New York to Syria, over 5,600 miles, and in 1937 a crew of three Russians flew an Antonov monoplane 6,300 miles from Moscow to California.

If it was desirable to fly or to have flown the Atlantic, or the Pacific, or around the world at all, then it was also desirable, once the challenge had been disposed of, to re-fly those flights in various combinations of numbers, directions, types of aircraft, sexes, and so on. Elsie Mackay, a pilot and the daughter of an English peer, wishing to be the first woman to cross the Atlantic in an airplane, in 1928 engaged a one-eyed airline pilot, Captain W. R. Hinchliffe, to accompany her across. She offered him a fee of 10,000 pounds. They disappeared, and despite mediumistic communications between Mrs. Hinchliffe and her late husband regarding the manner and place of their ditching, they were never found. The loss of her husband proved profitable for that lady, however, since Miss Mackay's father gave Mrs. Hinchliffe the 10,000 pounds instead.

Apart from the honor of being the first woman across, Miss Mackay may also have aspired to be aboard the first nonstop flight across the Atlantic from east to west. The first lady ambitious of that goal to lose her life in its pursuit was the sixty-one-year old Princess Löwenstein-Wertheim, who perished with two Englishmen in a 1927 attempt. Also in 1927, one Ruth Elder engaged a pilot to fly her from New York to Paris; they got lost and ditched near the Azores, alongside a steamer. They had flown long enough and far enough to reach Europe, but in the wrong direction, or directions. Nevertheless, their debacle attracted enough attention to get Miss Elder a film contract. Then a niece of the late President Wilson, Mrs. F. W. Grayson, engaged a crew to fly her from Maine to Denmark—a novel choice of termini; but after one abortive attempt and some shuffling of crew members, she took off on December 23 into dangerous weather, and predictably vanished soon after.

In 1928 Mrs. Frederick Guest engaged Amelia Earhart, a pilot with five hundred hours' experience and a physical resemblance to Charles Lindbergh, to be flown by Wilmer Stultz, a dropout from the first Grayson attempt, from Newfoundland to England in a Fokker trimotor. The flight was successful, and Amelia Earhart duly became famous; in 1932 she made her own solo Atlantic crossing in a Lockheed Vega.

As the thirties grew older, the crossing of the Atlantic grew easier; in 1933 twenty-five Italian airplanes did it in formation. The new challenge was the flight round the world. This had already been done in 1924 by a group of four Douglas biplanes with Army crews who spent half a year at the task. But it had to be done faster, if the practicality of the airplane for long-distance travel were to be convincingly demonstrated.

One of the major obstacles to a round-the-world flight was the lack of facilities for landing, fueling, or repairing airplanes in many parts of the world. When Charles and Anne Lindbergh flew from New York to China via the great circle in 1931, a flight beautifully recorded by Anne in her little book *North to the Orient,* fuel had to be shipped ahead well in advance to remote points in Canada, and any major mechanical breakdown would have spelled the end of the trip. On the other hand, the political barriers that

The ultimate hero of long-distance flying, and his airplane. Courtesy National Air and Space Museum, Smithsonian Institution.

have existed since 1945 had then not yet been erected, and the Lindberghs could stop in Kamchatka, and Wiley Post in Siberia, without difficulty.

It is almost forgotten now that the airplane was preceded in the conquest of distance by the airship. While primitive airplanes struggled to push outward the frontiers of their progress, the daring and stamina of their crews compensating for the inefficiencies of the equipment, giant dirigibles were already capable of astounding flights. Ballooning had a long head start on heavier-than-air flying. A lamb, a rooster, and a duck were the first sentient beings to ride aloft in a balloon, and though the lamb kicked the rooster before takeoff and broke its wing, they landed safely and were all probably later eaten for their pains. The first human balloon ascensions were made by Jean-François Pilâtre de Rozier and his friend, a young officer named d'Arlandes, who crossed Paris in a hot-air balloon in 1783.

D'Arlandes' account of the flight returns again and again to what was to be a recurrent problem of the early aeronauts: he became so fascinated by the spectacle of Paris seen from the air that he kept forgetting to feed the fire of straw which kept the contrivance aloft. An American surgeon, John Jefferies, who set out to make meteorological observations during a 1785 flight in England, forgot all about them, so mesmerized was he by the grandeur of the view. Ballooning made rapid progress; it was only ten days after the crossing of Paris by Pilâtre de Rozier that he and d'Arlandes ascended in a hydrogen balloon to the prodigious height of 10,000 feet. One can imagine their emotions. Men had stood in the mountains and looked out upon the plains before that, but never upon Paris.

Given the long history of the balloon, it is not surprising that in 1929, when ocean crossings in airplanes were still the province of a few daredevils, Dr. Hugo von Eckener, who had made several transatlantic flights in the giant dirigible *Graf Zeppelin,* announced that he would undertake a round the world flight, and offered luxury passage to twenty paying passengers at $2,500 a piece. The *Graf Zeppelin* was 776 feet long, beautifully streamlined, propelled by five engines of 550 horsepower each, and capable of a cruising speed of 70 mph. It compared quite favorably with the airplanes of the day, particularly in its accommodations: in addition to its crew of forty-one, it could carry twenty passengers in ten staterooms, and provide them with the advantages of three toilets, a galley, and a "social hall"-cum-dining room besides. Like most of the airplanes of the time it flew at low altitude, providing its passengers with an intimate view of the ground and observers below with the incredible spectacle of a flying object the size of an ocean liner eclipsing the sky. Over Russia, during the Germany-to-Japan leg of the flight, Karl von Wiegand, the Hearst papers' correspondent on the flight, wrote that the *Graf* "began to panic the ignorant peasants beneath us. ... Many of the peasants, who have never seen a train, evidently mistake the Zeppelin, with its thrumming motors and sylph-like speed and lines, for a

Amelia Earhart made a career of being a kind of female Lindbergh; she vanished during a transpacific flight. Courtesy National Air and Space Museum, Smithsonian Institution.

celestial monster. We saw villagers run wildly into forests and houses and gather around churches, gazing into the sky in awe and terror.''

Many of those primitive muzhiks probably never knew what they had seen. But the magnitude of the trip was not lost on the newspaper-reading world. Flying enormous legs—New York to Germany to Japan to California—which lasted days at a stretch, while carrying paying passengers in relative comfort, the *Graf Zeppelin* made the first decisive step toward replacing land and sea travel with air travel. There was still the mystery, the awe, the novelty and the adventure: one passenger wrote, as the *Graf* crossed Siberia, ''There were some trails, a few teepee tents, shaggy people, shaggy horses and wolf-like dogs, a few birds. In the water we could see, occasionally, large fish near the surface.'' There was an intimacy with the world in those early flights which was soon to be lost as the advantages of altitude were seized upon by pressurized airplanes with supercharged engines.

The airship, remarkable as were its capabilities, was not destined to inherit the earth. Ultimately the superior speed of the airplane would triumph; but airships were dealt a premature blow, perhaps, by the disaster of the *Hindenburg*. They may yet return, filled with inert helium rather than explosive hydrogen, powered by efficient, quiet modern engines, and capable of carrying huge weights economically at modest but adequate speed. They might find a place in the huge gap between ships and airliners, if their vulnerability to weather can be reduced.

When World War II came it brought with it the same sort of jump in the capabilities of airplanes that World War I had. In 1946 a Lockheed P-2V navy patrol plane flew nonstop from Perth, Australia to Columbus, Ohio, a distance of 11,325 miles. The present absolute distance record was set in 1962, by a Boeing B-52 which flew from Okinawa to Madrid—12,532 miles. But just as in the past, the capabilities of large military aircraft, developed during wartime, had soon filtered down to small civil airplanes, and the smaller products of the postwar era were capable, at least in principle, of the same feats as the P-2Vs and B-52s. A flight of more than 13,000 miles by a small civil aircraft is quite conceivable; the well-known promoter James Bede even built an airplane which was intended to circle the globe nonstop in a flight of a week's duration. It never made the flight; there were problems of high oil consumption at low power settings in the airplane's 210-horsepower Continental engine, and larger business problems which eventually compelled Bede to sell the plane. But there are no absolute obstacles to such a flight being made.

On a more accessible level, and one with more practical application, transatlantic and transpacific flights are easily within the capabilities of ordinary civil aircraft equipped with supplementary tanks. As Lindbergh and

While airplanes were still in their adolescence, huge airships like the 776-foot Graf Zeppelin *could carry passengers in comfort around the world, making only three stops en route*. Courtesy National Air and Space Museum, Smithsonian Institution.

many others after him demonstrated, it doesn't take a big airplane or a big crew to fly long distances. With modern equipment, flights that would have captured the attention of the world in the thirties are routinely possible. The world no longer cares, but for the pilots who make them, those flights represent the same challenge and yield the same aesthetic satisfaction as they did for the pioneers. To look at a globe and imagine one's long, curving track upon it, the slow progress of a speck across oceans and continents, to meditate upon the loneliness and the risk, produces in the mind an impression of almost cosmic grandeur and awe. It is that impression that fueled the conquest of distance in aviation's infancy, and the same impression that keeps it alive today.

Though we now take for granted a compact and portable Earth, presented to us in satellite photographs in the morning weather report, some part of us still nourishes an atavistic and wishful belief in the vastness of distances and in the inaccessibility, and the unique and unimaginable qualities, of remote lands. It is to gratify that belief that we undertake long flights; and it is because of it that we still experience a profound and memorable sense of happiness and wonderment when, after many hours of flying over open ocean, we discern on the horizon the dark unwavering line of a distant coast. There is a frightening solitude in the open sea; in crossing it we take the role of the mythical heroes who dove to the bottom of black lakes to battle a monster or flew up to pluck a star from the sky. The empty space between the continents is filled, like outer space, with the primordial unconscious of our species. We must experience it to understand it. The airplane has come close to destroying that reservoir of illuminating darkness. Airliners make the great and hostile gulfs among worlds seem trivial and insignificant.

But small airplanes do not inspire so much confidence; and in them we can still feel some of what the first pilots felt who left cities and shores behind them for the ordeal of a long-distance flight.

Every flight of any length, even far from oceans and crossing no national borders, contains the same elements. It seems to me tremendously satisfying to leave the earth in one place and come down in another, one recognizably different, remote, in another time zone or another latitude, with a different climate, landscape, and accent. These shifts of locale hold the essential pleasure of flying long distances, which continues to exist even though the superficial novelty of long flights has worn off, and they make no heroes any more. Although the era of famous long-distance flights is over, that of long-distance flying in general is only now coming into its own, with the ready accessibility of suitable equipment, the multiplication of navigational aids, and the easy availability of information about distant and foreign places. Unfortunately, at the same time that long-distance flights seem most attractive, they are made more difficult by the rising price of fuel and, in some

The current holder of the absolute distance record, a U.S. Air Force Boeing B-52. Its long, slender wings suggest an airplane optimized for long range. Courtesy National Air and Space Museum, Smithsonian Institution.

places, by political uncertainties. None of these obstacles is likely to disappear, but none of them is absolute, either. For the purely geographic difficulties, all that is needed is a certain recklessness—a willingness to trust everything to a roll of the dice, with the comfort of knowing that the odds are far more favorable than at the crap table. For the political ones, you must shed your attachment to plans and be prepared for a setback or an irritation. Obviously flights in distant countries cannot be conducted with the same reliability and fidelity to a schedule as those within the United States. Failure of an alternator may be a more irksome difficulty in Peru than in Iowa. You have to be prepared to take your chances.

When you study the globe with the thought of conquering it with an airplane, it surprises you. From Fairbanks to Miami, for instance, is only 4,000 miles, though the two places sound worlds apart. On the other hand, Rio de Janeiro is farther from Los Angeles than Tokyo is. The famous 3,000 miles from coast to coast in the U.S.A., a figure known to every schoolchild, is only 2,400 miles (L.A. to New York).

I was surprised, when I began to plan my own first ocean crossing, to learn that the distance from one side of the Atlantic to the other was less than 2,000 miles; it is about the same distance as from Los Angeles (where I live) to Detroit or Atlanta. World War II ferry and transport operations had produced airports at Gander, Goose, and Shannon, as well as at several places in Greenland and Iceland. The flight was tailor-made for a light airplane. Mine was designed to cover distances of more than 3,000 miles without refueling, and so there would be no reason to call at Greenland or Iceland. I had not made flights of eleven hours before, but between that and a flight of seven or eight, which I had made, there was not much difference.

I spent a great deal of time collecting information about the route. Some of it was superfluous. For instance, I called a ferry service in Wichita, Flo-Air, and a pilot there kindly read off to me the lat/long coordinates of the route they use. At the time I thought that this was valuable information, because it seemed consecrated by use, whereas if I had drawn a line on a globe and taken off the coordinates myself, they would have seemed speculative and somehow less dependable than those provided by a genuine ferry service. Of course, there was really no difference. Flying over water is no different from flying over land, so long as all goes well; a great circle is a great circle; and deviations of a minute or two, or a degree or two, this way or that make little or no difference in the final outcome. I had been impressed by the well-known story of Lindbergh's working out his route on a chart bought from a shipchandler, and of the uncanny accuracy of his landfall at Ireland. I did not understand until I had made the trip myself that the accuracy with which one could actually navigate was so much inferior to that with which one could plan that precision was completely meaningless, and that the exactitude of his landfall was entirely accidental. He was, after all, Lucky Lindy.

All flying involves emotional experiences which arise from the characteristics of the points of origin and destination, the quality of the air en route, the weather at the origin and the destination, the feel of the airplane, and of course your own moods and feelings independent of outside influences. After planning an Atlantic crossing for months, you arrive at Gander with a heightened sense of anticipation and dread—elements always mixed, I find, in ocean flying. The bulk of your trip obliterates everything else from your mental horizons, while at the same time you feel insignificant and unimportant amid the activities at Gander. There are other planes coming and going, jets, passengers, pilots, cargo, all the paraphernalia of a big airport, and your plane is just that little white thing far away on the ramp, and your weather folder is lost among the others. You feel as though people should look at you with concern, even awe, put down what they are doing, say, "Ah, really, well! . . ." But none of these things happens, and you are processed, along with your fantasies, routinely.

We took off, my friend Nancy Salter and I, at nine in the morning; there was a low fog hanging over the runway, and we plunged into it seconds after breaking ground. We had never taken off with a full load of fuel—nearly half a ton of gasoline in *Melmoth*'s little 23-foot wings—and it was an irrevocable test hop with no prospect, because of the fog, of coming back to land if the plane wouldn't climb. But it was supposed to climb, according to my calculations and my instincts, and climb it did. We broke out of the fog after several thousand feet over a blue sky, and after 45 minutes leveled out at 9,000 feet.

Soon the fog broke up beneath us, exposing the calm water, its gemlike blue separated from the powder blue of the sky by an indistinct horizon. The land was already far out of reach.

Although on later overwater flights you give up thinking about the destination—an activity which only makes the trip seem longer—on the first, you can think of little else but the relative size of the portions behind and before you, and of the gradual shift of your attachments from America (in this case) to Europe. The hours pass, one after another. It takes only a moment to say, "An hour passed." For it to pass, in real life, seems, when you are in the middle of the Atlantic Ocean, to take much more than an hour. And every hour is followed by another, and another. When you arrive in Ireland, it is visibly a different place than the one you left, visibly *Irish,* with its green hills and hedgerows, scattered villages and meandering roads. Everything that greets you—the countryside below, the accent of the controllers, the currency in which the landing-fee bill is written up, the side of the road on which the taxi drives, the tea at the Shannon Shamrock, the texture of the bed linen, the shape of the bathroom fixtures—acclaims your accomplishment and rejoices in your safety like a welcoming crowd.

Inevitably I have found the arrival delightful and the departure unpleasant, and so for the one have put on a relaxed and meditative frame of mind, attentive to every event and image, and for the other a hurried, distracted one, fixing my attention on routine details and getting aboard the airplane and into the air with a minimum of reflection.

I used to sky dive, and I noticed there, too, that the preparation for the jump, the climb to altitude, the stepping out of the airplane into space, and the descent and arrival upon the ground, all these steps were arranged, also, in order of agreeableness; the first steps had to be ignored so that the last could be savored. An ocean crossing is the same way.

Sometimes events pick you up and carry you; sometimes your mood permits you to take risks which under different circumstances you would abhor. There was a headwind of forty to fifty knots for much of our return trip from Ireland to North America, and I elected to fly via Iceland, although we could in principle have made the trip back from Shannon to Gander despite the wind. I had been aware of an increasing roughness in the engine on startup, which seemed to disappear after it had been running for a while. In Iceland, the roughness became severe. Suspecting a fouled plug, I found tools and shelter in a huge hangar from the storm that was raging outside, and pulled the plugs. There was no fouling. But I was so childishly impatient to leave Iceland, which seemed that day a tremendously dismal and inhospitable place, that I buttoned the engine back up and took off in spite of the uncertainty. The problem later turned out to be a bit of lint in an injector. The other problem was a harder one to deal with: my own impatience, readiness to ignore a danger signal, obstinacy, insistence on adhering to a plan. Perhaps if we had stayed a day in Iceland the storm would have ended, and we would have had a day of beautiful weather in which to visit what by all accounts is an interesting place. But at the time I couldn't think that way. I was keyed up, had to keep moving, like those souls driven through Dante's hell by a whirlwind.

A year later the roles were reversed: circumstances had taken the initiative, and I had become passive. Nancy and I were in Cold Bay, Alaska. The plan was to stay overnight there and take off for Japan, 2,500 miles away, the next morning. But a powerful weather system, the remnant of a typhoon, was moving up the central Pacific toward the Aleutians and would arrive during the night. Its counterclockwise flow would give us a strong tailwind as long as the low center was to the south and east of our course; once it crossed our track, it would provide a headwind that would put Japan out of our reach. So without resting—it was nine-thirty in the evening, we had just flown five hours from Anchorage—we refueled and took off for a night flight of nearly fifteen hours to Chitose, in Japan.

I cannot think in what important way those flights were different from those of the twenties and thirties. They were undoubtedly safer, in every identifiable sense. But the pilots and the passengers, then as now, do not spend their time working out the odds. The idea of danger is repressed; it must be repressed, or the flight would be misery. It is only the seasoning of the feast. The rest is the *going,* the sense of remoteness overcome, the strange places, and the dueling with chance. None of that has changed; it is there for the taking today, as it was when Alcock and Brown crossed the Atlantic, or Charles Kingsford-Smith the Pacific. It is the essence of airplanes, and the unchanging romance of flight.

2

MODERN EQUIPMENT

Today's light aircraft are capable of longer and more difficult flights than many of their pilots are. If three or four hours was considered an adequate endurance in the forties and fifties, five or six hours, or in some cases seven or eight, is more typical today. The magic figure, in range specifications, is 1,000 nm. Twins generally have somewhat greater endurance than singles, presumably because they are thought to be more comfortable, more likely to be equipped with automatic pilots, and so more tolerable for flights of six hours or more. The Grumman Cougar boasts an endurance of 8.7 hours at 65 percent (no reserve) for a gross still-air range of nearly 1,300 nm. The majority of singles and light twins of 200 hp (horsepower) or more can make or approach the 1,000 nm figure.

At the same time they are easy to fly, quiet compared with earlier generations of private planes (though not when compared with almost anything else), and often equipped with at least a simple autopilot, so that pilot workload is small and long flights are not necessarily very taxing. They are reasonably well heated and ventilated and do not vibrate excessively. Highly automated radio equipment presents navigation information in a readily digestible form; the dissemination of en route information is efficient; charts are of good quality; communications are clear and reliable. Compared with the pilots of the twenties and thirties who made the pioneering long-distance flights, the pilots of today have their work done for them. They have only to supervise events which unfold almost automatically.

What is still pushed to its limits today, when airplanes are used to the maximum, is the pilot. If an ILS (Instrument Landing System) approach to minimums comes at the end of a six-hour flight, it is not the ILS receiver but the pilot that will show the wear and tear. Because it is so easy to tax one's endurance on really long flights, equipment plays an important role. The better the equipment, the easier the pilot's work.

Airplane operator's handbooks give range figures for different conditions of operation. In recent years there has been some standardization of formats, and in any case the conditions presupposed in determining range are always clearly explained. Normally, a forty-five-minute reserve is included, because it is the legal minimum IFR reserve; furthermore, only so-called usable fuel is considered. Usable fuel is fuel which will feed to the engine under certain specified conditions of slip or skid, climb or descent; unusable fuel may be usable under some conditions of flight, but not under others, and so it is omitted from consideration.

By comparing ranges with the power settings at which they are attained, you can find the average power setting for maximum range. This will be an approximate figure, since best range is actually achieved by gradually reducing the power setting as fuel burns away; but as the range charts will show, the difference in range between the best power setting and one a few percent off the best is very small—ten or twenty miles, in many cases. By cross-checking the power setting for best range against the speed/power charts for several altitudes, you can find the most efficient indicated speed of the airplane.

Different manufacturers may arrive at their performance figures in different ways, and it is often puzzling to see two airplanes built by different manufacturers, with the same powerplant, giving different fuel flow figures for the same percentage of power. Furthermore, pilots often find that their own experience differs from book, either for better or for worse. The best way to get a handle on the gasoline mileage that can actually be expected from your airplane is to consult the power/speed/range information in the manual for the airplane's potential performance, and then to keep a log of time flown, fuel burned, and speeds actually encountered, for fifty or a hundred hours. Airspeed indicators are not always very accurate, and a rec-

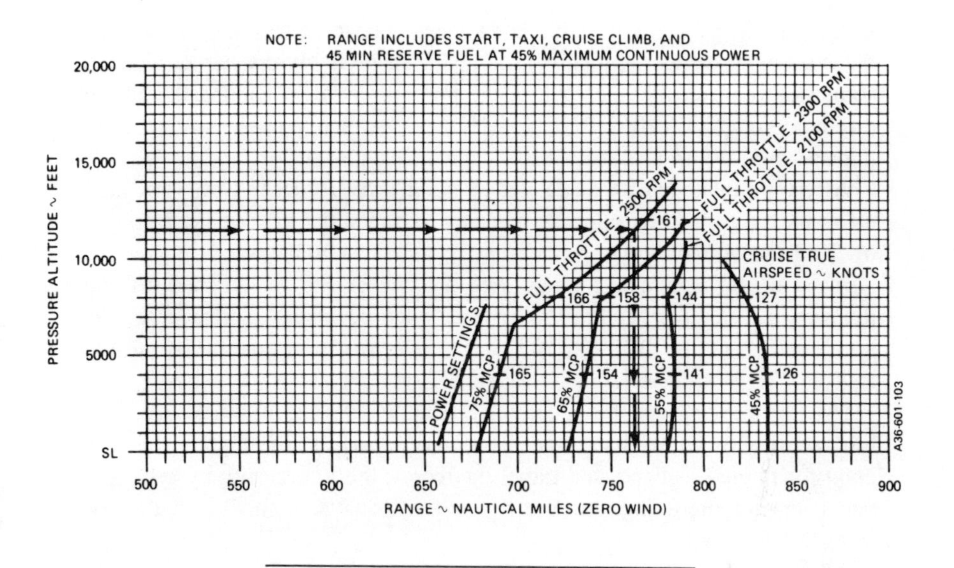

Pilots' handbooks give a rough idea of the power setting for maximum range.

ord of actual block speeds, fuel burns, and oil consumption on a number of trips, although it does not have the theoretical purity of handbook optimums, is probably more useful in the long run. Bear in mind that the average fuel burn for short trips will be higher than that for long trips, because the fuel used in taxiing, taking off, and climbing will count for proportionately more.

For long flights the single most useful aid, besides ample fuel tankage, is an autopilot. The simplest autopilot is a wing leveler, and at least some wing levelers (such as the Mitchell Century I) can be hooked up to VOR outputs in order to track omni radials or ILSs. In order to be useful for more than emergencies, a wing leveler must have a roll trim. Carefully trimmed, it will hold heading just as reliably as a coupled heading hold, which represents the next level of sophistication in autopilots, and costs considerably more than the plain wing leveler. Heading hold capability enables the autopilot to keep on a heading selected on the DG (Directional Gyro), or to fly "command turns" to selected headings. The selecting is done by means of a rotatable "bug" on the periphery of the DG dial. Since DGs drift at a rate of several degrees every half-hour or so, the heading hold is not a perfect solution to the problem of maintaining a steady heading for long periods and the plain wing leveler, properly trimmed, can probably do just as well. A somewhat more exotic approach to the problem involves slaving the DG to a remote compass (called a "flux gate") so that it is not necessary for the pilot periodically to reset the DG.

Airways flights can be made with a wing leveler/tracker with very little effort on the part of the pilot; he has only to remember to keep the autopilot tracking the correct VOR. Sometimes waviness in VOR radials makes the airplane weave from side to side; then the tracker can be switched off and the wing leveler (or heading hold) function used alone. In that case, the pilot must check the OBS (Omni Bearing Selector) from time to time to see to it that he is on course.

Usually tracking can be accomplished at two levels of sensitivity. The high sensitivity setting holds the plane very closely to the radial, and it is good for flying localizers and radial intercepts; the low sensitivity setting is better for en route flying, because it is less responsive to small irregularities in the VOR signal.

Altitude hold is the last thing to be added to an autopilot (other than, on large aircraft, yaw damping), and it is the least necessary for light aircraft flying. Light airplanes hold altitude quite well, hands off, for long periods.

In the area of avionics there are several conveniences that make long-

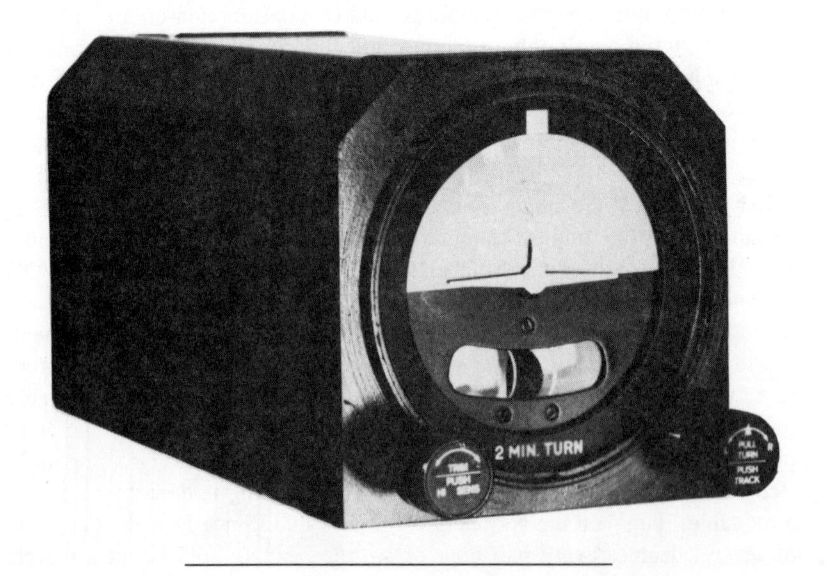

An autopilot, even one as simple as this Mitchell Century I "wing leveler," is a tremendous help on long flights. Courtesy Edo-Aire Mitchell.

distance flying easier. One of the most useful is DME (Distance Measuring Equipment), assuming that an airplane is equipped with VOR and ADF (Automatic Direction Finding) receivers to begin with. For reasons which I will explain in another chapter, DME is more than a mere navigational tool; it makes possible faster, more efficient flights because it gives the pilot access to instant groundspeed information. Since VOR, DME, and ILS/localizer functions can be combined in a single radio to provide Rnav capability as well, as they have been in the King KNS 80 and undoubtedly will be in other companies' products, the price of including DME in the panel, which has always been comparatively high (beginning around $3,000), will be coming down (at least in relative terms).

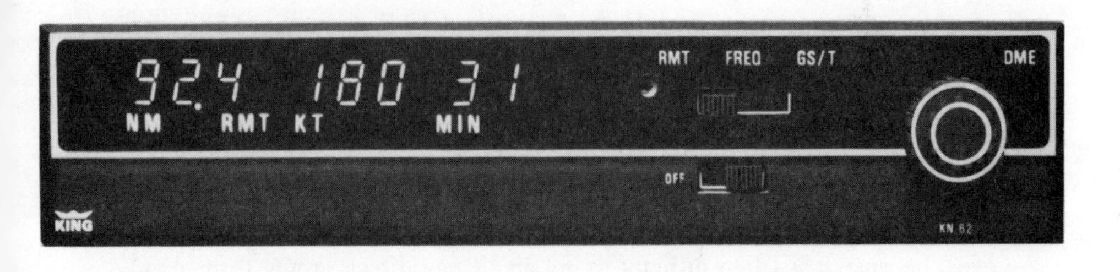

DME is a relatively expensive ($3,000) but very useful gadget, both for general navigation and for en route estimation of winds. This one is King's highly miniaturized KN 62.

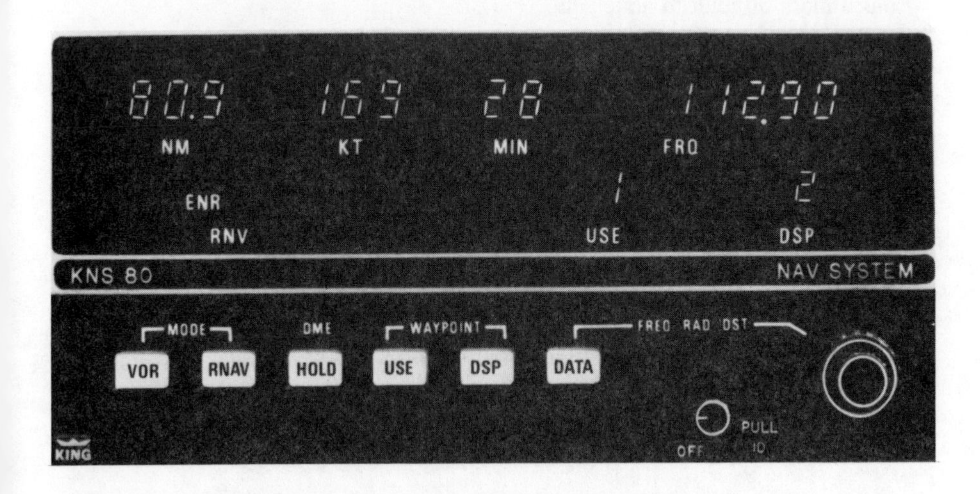

King's KNS 80 has VOR, DME, and Rnav integrated into a single box; most Rnav sets are separate units which receive information from independent VOR and DME receivers.

Particularly in VFR weather, navigation can be made easier and more efficient by an Rnav (area navigation) computer. An Rnav set synthesizes an imaginary VOR using information triangulated from other VORs. The pilot is able to place this imaginary VOR wherever he wants it. Thus, the OBS can be used to fly any track at all, not just straight lines between omni stations. By putting the imaginary VOR on an airport which he is trying to locate in conditions of poor visibility, the pilot can fly to that airport as though it had its own VOR. Trying to make a VFR penetration of an area of bad weather, he can place a VOR in an area of good weather to which he might want to return, and get back there without having to keep constant track of his position while scud-running.

Rnav only works when two VORs, or one VOR with DME, can be received; the imaginary VOR is liable to disappear when one gets too low, especially in the west, where omnis can be very far apart and the terrain irregular. Apart from its usefulness in low-level VFR orientation, however, Rnav can be used to trace a straight track from start to destination, without the detours and doglegs that the airways system sometimes imposes. This ability is usually more significant on short trips than on long ones, but even on long trips some savings in time and fuel can be achieved by flying straight tracks.

The ability to fly direct is more useful in VFR than in IFR flying, however, because it is often difficult to get an "Rnav direct" route from ATC. The reason is quite obvious. So long as all airplanes are following the existing route system at alternating altitudes, potential conflicts are easily foreseen by controllers. As soon as a significant number of airplanes begin to fly paths crisscrossing one another and the airways system at random, however, the potential number of conflicts rises sharply, and they become much more difficult to anticipate.

Rnav direct routings can be had most readily in the western states, where traffic densities are lower than in the east, and at altitudes between 10,000 and 18,000 feet, where there is relatively little traffic, and most of it is, if not on an IFR flight plan, at least equipped with an encoding altimeter and squawking Mode C while VFR. It is also frequently easier to get Rnav direct routings by requesting them en route, for instance to avoid a dogleg on an airway. "Request direct So-and-so if possible" gets an approval if traffic at the moment permits, whereas an Rnav direct flight plan is likely to be disapproved, and an airways route substituted, merely as a matter of principle.

Although many instrument approaches, especially at smaller airports, use ADF, it is not extremely useful as an en route aid. At best, it can occasionally be used to verify progress along a track that passes by an NDB, or to set a waypoint between two VORs when there is no airway. However, as

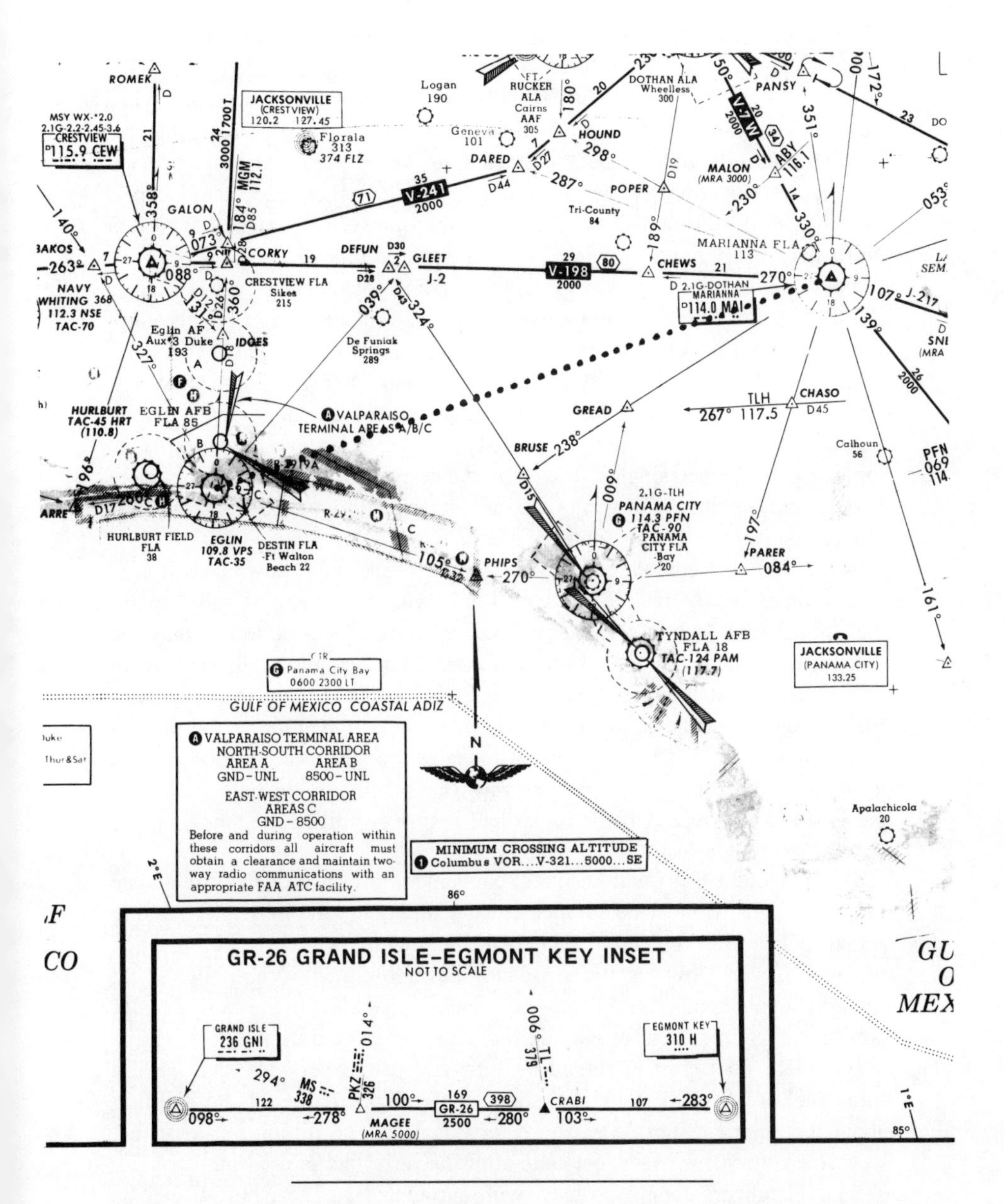

Over long distances, Rnav straightens out the bends in the airways system. It also overcomes problems in locating places that don't have co-located or nearby radio aids. (Map not to be used for navigational purposes).

soon as you leave the United States ADF grows rapidly in importance. On ocean routes it is indispensable, not only because of the prevalence of NDBs over VORs in most areas, but also because of its much longer range. When you are navigating toward a small island in a large ocean, anything that increases its effective diameter is welcome. Omni seems to propagate unusually well over water, reaching out 150 miles when one would expect over land to lose it at 100 miles or less; but a powerful broadcast station or NDB may easily have a range of several hundred miles.

For long-distance flights, especially where supplementary fuel is used, a digital fuel flow indicator and a totalizer are extremely handy, though, since they are quite expensive, most pilots make do without them. The commonest kind of fuel quantity transmitter in light aircraft is the float-type potentiometer—the same kind of device as is used on cars—and it is neither overly accurate nor perfectly reliable. More expensive airplanes, starting with the bigger singles, use capacitance-type senders; these are more precise and, having no moving parts, are supposedly more reliable, but they are still capable of giving spurious readings.

A totalizer is a computer combined with a clock and a flowmeter; it constantly measures the rate at which fuel is flowing to the engine and displays that information, while also integrating fuel flow over time to compute the amount of fuel used. Having been programmed by the pilot with the amount of fuel aboard at the beginning of the flight, it can display the amount of fuel remaining in the airplane and the flight time remaining at the present fuel-flow rate. Like all computers, it looks as though it knows more than it does; the totalizer would have no way of knowing if the pilot had programmed it incorrectly, if the fueler had improperly filled the tanks, or if a tank had sprung a leak.

While a totalizer gives immediate, convenient information about fuel state during a flight, it must be supplemented with some kind of sender reading actual fuel level in the tank for a dependable picture of fuel remaining when the fuel is very low. A few airplanes have low-fuel warning lights; all should. Supplementary tanks (or "ferry tanks") installed in the cabin usually don't have any means of displaying fuel level, unless a transparent standpipe is provided. Fuel burn is judged by time and past experience, and usually some fuel is left in the tank at the end of the flight. If there is no question about the engine restarting easily, a ferry tank can be run dry; at least that way you know your exact fuel state after the ferry fuel is used up.

Some totalizers cannot be used with carbureted engines, but all can be used with any kind of fuel injection system, whether it is one that returns fuel to tanks (like Continental's) or not (like the Bendix system).

Especially for overwater flights where overload fuel is carried and where

A fuel flow totalizing system permits absolutely accurate knowledge of fuel state. Courtesy Silver Instruments, Inc.

it may be desirable to fly at the most efficient speed, an angle-of-attack indicator can be useful. Most of the important speeds of the airplane—stalling speed, speed for best rate or angle of climb, speed for best range—increase with increasing weight, but correspond to a constant angle of attack. This is the virtue of angle of attack as a parameter of flight: it is independent of weight. Most jet aircraft and military aircraft use angle of attack information as a primary reference during the approach; carrier approaches rely on angle of attack exclusively.

There is one angle-of-attack indicator, or "lift indicator," on the market that is suitable for light aircraft; it is made by Safe Flight, of White Plains, New York, and it mounts in the leading edge of the wing like a stall warning tab, which it resembles. Although it is intended as an aid for approach and initial climb, one can by experiment discover the indication corresponding to the most efficient speed, and set up a cruising speed using the lift indicator rather than the airspeed. In landing and taking off with unusual loads, the angle of attack indicator automatically signals the correct attitude to fly, regardless of weight.

Certain other kinds of instrumentation, normally omitted from light aircraft, are very desirable for long-distance flights, especially over water. For instance, a voltage meter is useful both to monitor the condition of the electrical system and, in case of an alternator failure, to keep track of the amount of power remaining in the battery. A low-oil-pressure annunciator is desirable, because certain kinds of malfunctions, such as broken piston rings, produce increased oil consumption and could run an engine out of oil on a long flight, although they would not be likely to on a short one.

For reasons which I will go into in more detail later, cylinder head temperature (CHT) and exhaust-gas temperature (EGT) instrumentation is an aid to efficient flying. It is best to have a readout from each cylinder, not from only one. I have had good service from a twin-engine CHT gauge in my single-engine plane; using a double-pole, triple-throw rotary switch, I can display CHT from the front, middle, or rear pair of cylinders. Various EGT equipment is also available; I favor the type, such as the Alcor, which has a narrow band of temperatures and large 25-degree intervals. The type of gauge that displays a wide temperature range does not permit fine tuning of the mixture.

I have also installed in my airplane a twin-engine manifold pressure gauge. One needle is plumbed to the intake manifold, the other to the static system. My reasoning is that the static needle serves as a crude backup to the altimeter, reading about 30 at sea level, and losing about one inch per 1,000 feet. In case of an altimeter failure, I would have in that needle some indication of my altitude, plus or minus 200 feet or so. Any manifold pres-

The Safe Flight angle-of-attack indicating system not only compensates automatically for aircraft weight, but can be used to maintain most efficient cruise speed. The middle bar represents approach speed, 1.3 V_s. The most efficient speed can be found experimentally at one weight, and then flown at any weight; it will be somewhere around the lower diamond. Courtesy Safe Flight Instrument Corp.

sure gauge can be used in this way, however, by firewalling the throttle and lowering the rpm as much as possible. The indicated full throttle manifold pressure will be half an inch to an inch below the ambient pressure, with the smaller errors at lower rpms.

It may seem pedantic to seek redundancy in things like altimetry, but altimeter failures are not unheard of, and altitude can be a very important piece of information, for instance if you have to make an instrument descent. Another way to get redundancy in altitude is through a blind encoder. A blind encoder is one which contains its own pressure sensor, rather than taking altitude information from the workings of the pilot's altimeter. Some are even available with a digital readout in the cockpit, which displays the same information as the encoder is sending down to the ground. Lacking this, in case of an altimeter failure in an area where ATC has Mode C capability, one could still get one's own altitude squawk back from ATC.

One area where redundancy is absolutely mandatory is in blind-flying instrumentation. Most airplanes have a DG and artificial horizon (AH) for primary reference, and a turn-and-bank and wet compass for backup. It is essential, in such a system, that the primary and backup instruments have separate power supplies. If the AH and DG are air-driven, the turn-and-bank or turn coordinator must be electric, and vice versa. Furthermore, since the DG is actually dependent on the wet compass for orientation over long periods, I would consider a second compass desirable for overwater flying out of range of navigational radio aids. Either a spare wet compass or a remote compass will do; the latter system involves a power supply, and is therefore dependent on the electrical system. If the DG is electric, and a remote compass is used as a backup for the wet compass, then it is important to swing the wet compass both with the airplane electrical system on and with it off, since if because of an electrical failure you are reduced to depending wholly on the wet compass, the electrical system will be off. If the compass had been swung only with the electrics on, it might be quite inaccurate.

Because of the role played by dead reckoning in much long-range flying, especially over water, swinging the compass before a trip is very important. In the United States one gets into the habit of using the compass and DG for general direction, and VOR radials for fine tuning; and if the two do not agree, one attributes the discrepancy to a cross-track wind. An accurate compass, in other words, is not too important when flying in the middle of a grid of navaids. When swinging the compass for an overwater trip, you should not only perform the operation with and without electrics, including HF (High Frequency comm radio), but also plot the deviations on a piece of graph paper. They should form a symmetrical sine curve about a median line. By smoothing the curve you can weed out errors in the swinging operation.

The second needle of a twin-engine manifold pressure gauge can be plumbed to serve as a standby altimeter. Note the close agreement between the lighter needle (about 9.4 spaces below 30) and the altimeter (9,580 feet).

Again, because dead reckoning is entirely a matter of the compass and the clock, the clock is as important as the compass. Because of the possibility of electrical failure, I would prefer a wind-up clock to an electric one, but in any case I would want a watch as well, preferably an alarm watch. On long overwater flights it may be possible to catch a little sleep, but one would not want to lose track of tank-switching or position-reporting time. An alarm clock would serve the same purpose.

A great many models of aircraft have lately been improving their performance with turbochargers. A turbocharger is an air pump driven by expanding exhaust gas. It compresses the thin ambient air of high altitudes to sea-level density or more, and prevents the engine from losing power with altitude, as normally aspirated engines inevitably must. Turbochargers have been in use for a long time on trucks, tractors, and automobiles, as well as on more sophisticated airplanes, and now that they are filtering down into comparatively inexpensive airplanes they come with a long history of service experience and a reputation for reliability. As one turbo manufacturer's sales representative put it, the turbos themselves have only three modes of operation: either A-OK, losing oil, or seized up. Their control systems, however, can be quite a bit more versatile.

Because they require an oil supply from the engine, with the attendant oil lines, fittings, check valves, and so on, and because they put much greater strains, in the form of pressure and heat, on the exhaust system than it would otherwise see, turbochargers take their toll in reliability and in cost of service. In exchange, they give a striking improvement in what could be called "vertical capability." A turbocharged airplane, confronting a line of weather with high buildups, can climb to 20,000 feet or more to cross, while a normally aspirated airplane would run out of climbing ability at a lower altitude and, just as important, would have a poor rate of climb from 10,000 feet upward. The turbocharger also gives some increase in cruising speed at high altitude and, under some conditions, improvements in efficiency (though the maximum efficiency of a turbocharged airplane is normally inferior to that of a normally aspirated one). It also makes the occasional very strong winds at high altitudes usable, if they happen to be tailwinds; and so a turbocharged airplane sometimes shows its pilot very gratifying groundspeeds. When the wind is a headwind, the turbocharged airplane is at any rate little worse off than a normally aspirated one would be.

When I say that a turbocharged airplane is normally less efficient than a normally aspirated one, I mean that in the strict scientific sense: all other things being equal, it makes worse use of its fuel. I will discuss the subject of fuel efficiency in detail in another chapter, but it is so widely believed that a turbocharger makes an airplane get better gas mileage (because it goes faster at the same power setting) that I should clarify here that a turbocharged airplane is *less* efficient at its *most* efficient speed; but it *may* be more efficient overall, because it *may* cruise more often at a higher altitude and a lower indicated airspeed, and therefore have a higher airframe efficiency, than a normally aspirated airplane. The gain in efficiency depends, however, on choices made by the pilot.

The most striking potential benefits from a turbocharger are in the altitudes above 15,000 feet. However, because they anticipate that most users will not want to depend on oxygen and will usually cruise at 12,000 feet or below, some manufacturers, such as Mooney and Piper, have deemphasized maximum high-altitude performance, settling for more modest improvements in cruise speed and rate of climb at more commonly used altitudes in exchange for simplicity of the turbo regulating system. An expression of the efficacy of a turbocharging system is its critical altitude; this is the highest altitude at which it is able to produce its rated manifold pressure, and it varies between 15,000 feet for the less effective systems to well over 20,000 feet for the most.

Oxygen is the hidden price of turbocharging. Not only is oxygen expensive and often impossible to get, but it puts its own limits on the duration of flight at high altitude.

Oxygen range is easily calculated. The capacity of oxygen bottles is given in cubic feet; this figure refers to the volume of oxygen *at sea-level pressure* that has been squeezed into the bottle. Sea-level pressure is 14.7 pounds per square inch (psi). The maximum pressure in normal bottles is 1,800 psi, although one can fill them to 2,000 psi quite safely, especially if the oxygen is to be used promptly and not left sitting in the sun. (Heat raises the pressure further.) The gas in the bottle is therefore compressed 1800/14.7, or about 122 times, and the actual volume of the bottle is 1/122 of its stated volume; in the case of a 38 cubic foot bottle, .3115 cubic feet.

Oxygen flow is controlled by a regulator. Most light aircraft use constant-flow systems which supply the user with a constant stream of pure oxygen; this is mixed with ambient air in a bag attached to the mask. A few, especially sailplanes, have the more sophisticated "demand" type system. With a constant-flow system, the regulator may simply provide oxygen at a certain steady pressure, or it may proportion the supply pressure to the ambient pressure. One needs about 1 liter per minute per 10,000 feet of altitude, beginning at 10,000 feet. Actually, no oxygen at all is normally required at 10,000 feet. At 15,000, however, 1.5 liters per minute is required; at 20,000, 2 liters; and so on.

A liter is 61.025 cubic inches; there are therefore 28.32 liters in a cubic foot, or 1,076 liters in a 38 cubic foot bottle. Filling to 2,000 psi, you can get nearly 1,200 liters into such a bottle.

Oxygen bottles should be filled slowly. The compression of gas inside the bottle releases heat, and if you fill a bottle to 2,000 pounds and then allow it to cool to ambient temperature, the pressure will drop noticeably. It will drop still more at a higher altitude, if the bottle is located in an unheated part of the airplane. Slow filling is thought to minimize these losses, but it is probably more practical, if you have the time, to fill the bottle, allow it to cool, and then top it off again.

The fee for filling an oxygen bottle is the same regardless of its size; it is not the oxygen but the labor that you are paying for. The cost of using oxygen is therefore inversely proportional to the size of your bottle; the larger the bottle, the less you pay per hour. Another plus for big bottles is the rarity of places where you can get refills. Most small airports simply can't do it; at the larger ones, not all FBOs (Fixed Base Operators) can. The best bet is to head for a place that sells turbine fuel and has a few jets or turboprops parked on its ramp. All high altitude airplanes, although they are pressurized, have standby oxygen systems. The price for a refill at such places tends to be high; $25, for instance, versus $10 at less posh (but harder to locate) establishments.

To determine oxygen range, you have to consider both bottle size and altitude of flight. If, for instance, you plan to fly at 17,500 feet with two people in the airplane, your combined need will be 3.5 liters per minute, and a 38 cubic foot bottle will theoretically keep you going for 5.71 hours. Looking at it another way, you could figure that you will take 9 hours to make a 1600 nm flight; the same bottle will provide 1200/9 or 133.3 liters per hour. That makes 2.22 liters per minute, or only 1.1 liters per minute per person—not enough to do any good, since it is the requirement for an altitude—11,000 feet—at which no oxygen is really needed at all. (Oxygen is *legally* required only at 12,500 feet and above.)

The rule of 1 liter per minute per 10,000 feet is only a simple rule of thumb, which you are at liberty to violate in the privacy of your cockpit; and the idea of an "oxygen requirement" is a hazy one, since different people's metabolisms differ, and the degree of mental deterioration owing to low oxygen uptake is hard to measure at low altitudes anyway.

At altitudes below 16,000 feet, nonsmokers in good physical condition may suffer some deterioration in mental acuity after an hour or more without supplementary oxygen; but cruising an airplane does not require much mental acuity. Although you cannot feel that you are hypoxic, and are likely to feel quite fit and even rather euphoric, you can view your own feelings with skepticism, and compare the length of time it takes you to determine your position on a chart, tune a radio, or give a position report with the time you would normally take. Sometimes hypoxia produces a lazy, tipsy feeling, such as might be brought on by a couple of glasses of wine. At any rate, you can tolerate mild hypoxia if you have oxygen available in case a stressful situation arises to require your full faculties; hypoxia vanishes immediately upon application of oxygen. This is equally true during the descent for landing. You may be as hypoxic as you like, and by the time you have descended to land you will have recovered.

I don't want to suggest, however, that hypoxia is not dangerous. It can be very dangerous, especially if it is not taken seriously; I have heard it said that more high-altitude bombers were lost in World War II to oxygen system malfunctions than to enemy fire. Severe hypoxia may bring on a feeling of well-being or indifference so great that the pilot does not take the trouble to lift his oxygen mask to his face, even though he sees his airplane going out of control. Perhaps such a sublime state of fearlessness is to be envied; but most of us have promises to keep, and would like to go a few more miles before we sleep.

Oxygen may affect visual acuity at night, especially in older people, and some recommendations on oxygen use are very conservative, suggesting at least intermittent oxygen above 5,000 feet at night and 8,000 feet during the day. On the other hand, I believe that during World War II the Navy used to send fighter pilots to their destinations at 16,000 feet without oxygen, and have them start using oxygen before entering combat. I hesitate to report this figure, of which I am uncertain, or for that matter the fact that I have often flown for longish periods at 14,500 without oxygen without noticing much deterioration in my performance (which is imperfect even at sea level) because there is a compulsive macho in many pilots, including me, which makes us apply the most extreme, least conservative rules to ourselves. I think that the FAA requirement of oxygen for the pilot for flights of more than half an hour at 12,500 feet is not excessively conservative, and probably should be adhered to.

A great many light-aircraft oxygen systems do not proportion flow to altitude; they simply flow, say, 2 liters per minute, and are "rated" for use at up to 20,000 feet. At all altitudes below 20,000 feet, they waste oxygen. A device called a Ted Nelson Oxygen Flow Meter (available in three models, depending on the type of system in the airplane, starting at $19.50 each, from Ted Nelson Co., 8638 Patterson Pass Road, Livermore, California 94550) allows the pilot to adjust his oxygen flow to precisely the FAA-approved level (or a lower level, if he sees fit). One is needed for each mask. At the very least, the Nelson flowmeter gives visual assurance that the oxygen system is functioning properly; at most, it can greatly reduce your oxygen consumption and increase your oxygen range.

Incidentally, it is pressure altitude, not density altitude, that determines the need for oxygen. You might suppose that if the cabin temperature is 70 degrees at a pressure altitude of 22,000 feet, the cabin density altitude is about 27,000 feet, and so there would be very little oxygen around. But what limits the ability of the body to absorb oxygen is not the amount in a

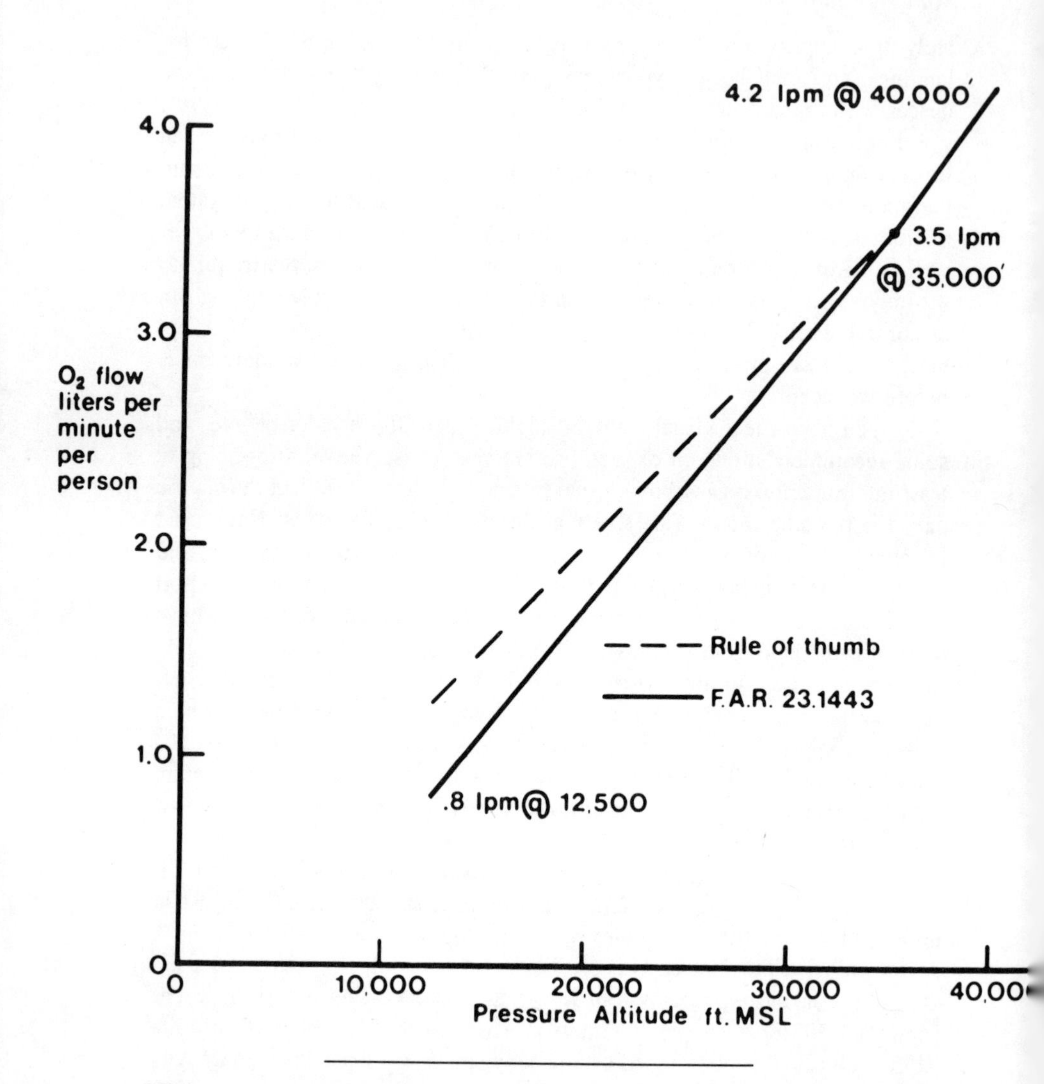

Minimum oxygen flow per person, according to the FAA. Up to 35,000 feet, the formula is: flow (in liters per minute) equals .12 times the altitude (in thousands of feet) minus 6.

cubic foot of air—you could always breathe a little more deeply or rapidly—but the "partial pressure" of the oxygen. Oxygen constitutes about a fifth of the gas in the atmosphere, and the "oxygen partial pressure" is the product of that fraction and the ambient pressure. In order for the membranes in the lungs to permit oxygen to pass into the bloodstream, the partial pressure must be above a certain value. Without pressurizing the airplane, you can't affect the ambient pressure; so instead you increase the fraction of the gas you are breathing that is oxygen, by feeding pure oxygen into a perforated face mask.

Occasionally somebody suggests running a tube from the turbocharger to the face mask, thus supplying compressed air to the pilot. This would not work. The face mask would simply be lifted from the pilot's face by the stream of air, which, by the way, would also be quite hot. If the pilot were to hold the air hose in his mouth, he would begin to inflate himself like a balloon. Air must be breathed at the ambient pressure surrounding the whole body; it cannot simply be supplied in compressed form at the nose or mouth.

It is possible to buy "solid oxygen" for emergency or short-term use. A solid oxygen system consists of cartridges of chemicals which burn slowly, releasing oxygen which may be breathed. A two-cartridge system weighs 3.5 pounds empty, and each cartridge weighs 1.5 pounds and lasts about a half an hour. The oxygen flow from each cartridge is 3 to 4 liters per minute. This is obviously excessive for a single person. By securing a tee to the kit's single outlet and then using Nelson flowmeters to ensure that both users are getting equal amounts of flow, you can use the system to serve two people at altitudes of 15,000 to 18,000 feet. So far as weight is concerned, the arrangement is a good one for an occasional long flight; at three pounds per one or two persons per hour (the flow cannot be regulated; the consumption rate is the same whether one or two people use the system), you could supply two people for five hours for about the same weight penalty that a conventional bottle of compressed oxygen would exact. The cost is another matter. At this writing, dry oxygen cartridges are selling for about $17 each. But oxygen refills are always quite expensive, and at any rate $170 for five hours of oxygen may not seem like a terribly heavy expense if compared with certain landing fees or fuel prices. On the other hand, if flying at oxygen altitudes for 5 hours were to pick you up a 20-knot tailwind, it would gain you in all only 100 nm. At 150 knots, you would be paying $4.25 a minute for the time you saved.

On long flights, there are comfort problems with oxygen. The mask, especially if it is heavy or fits tightly, may cut off circulation in parts of the face and cause increasing pain. Aviation oxygen is particularly dry, and some people complain of throat irritation after using it for a long time.

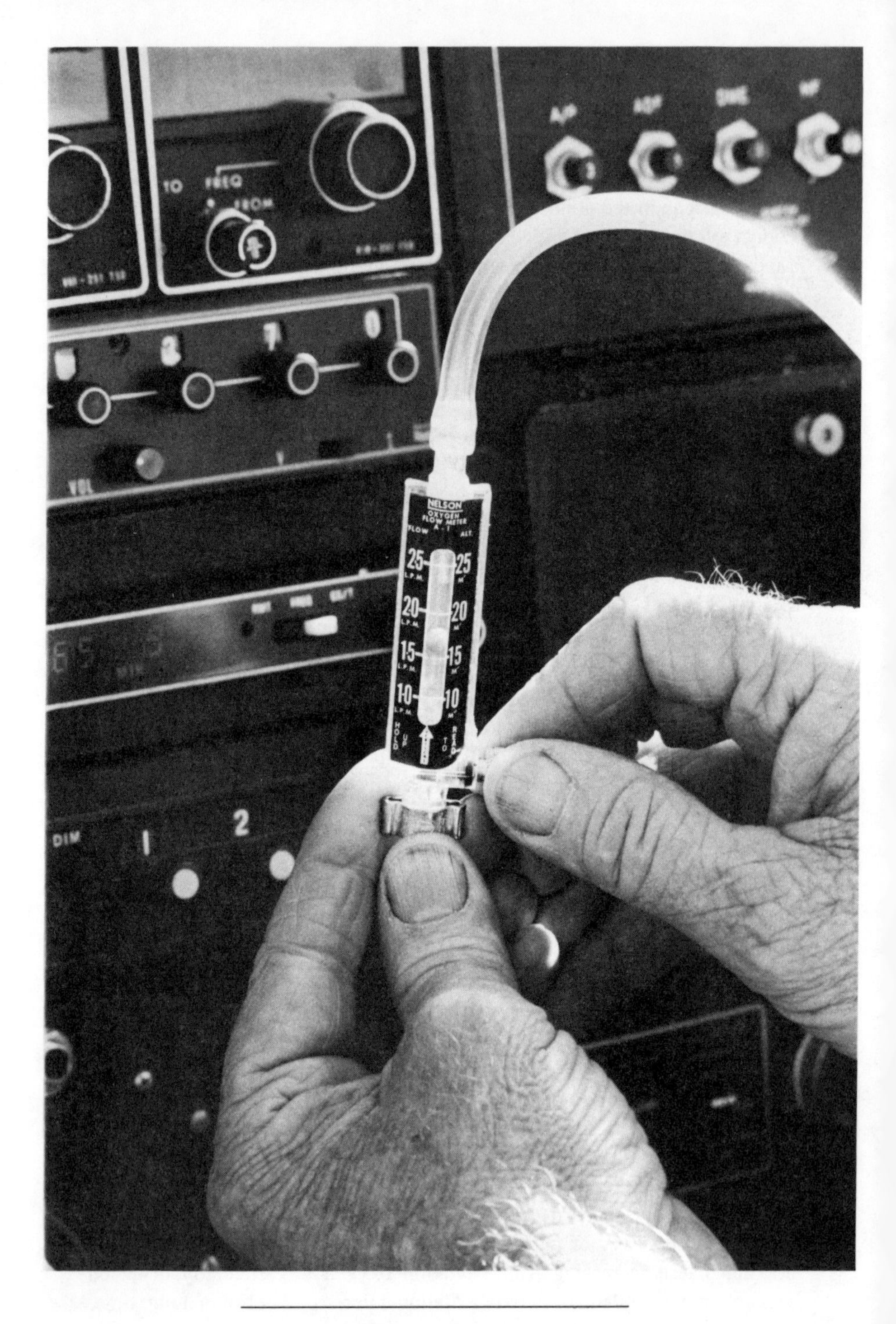

The Ted Nelson Oxygen Flow Meter.

Medical oxygen is wetter, and if the system is entirely within the cockpit there is not much chance of freezing (though there is a temperature drop as the oxygen leaves the bottle). But you can't get medical oxygen at airports. Of course, if oxygen were pleasant, like nitrous oxide, there would be no pressurized airplanes. It is an inconvenience you have to put up with in exchange for the speed you might get at high altitude.

There are two types of oxygen masks that are readily available, apart from all sorts of military and medical surplus equipment which you might come upon. For constant-flow systems, masks are always equipped with a "rebreather bag" which stores oxygen during exhalation and also, in some cases, mixes your breath with oxygen. One type, seen in most light aircraft "portable" systems, made by Scott, is of a light gauge of rubber, and it is held to the face by a 1/4-inch elastic band and a soft aluminum clip over the nose. The fit of these masks is imperfect, the elastic band eventually loses its grip on the mask, and the masks don't store well; if they get squashed and then get warm, or cold, or something, they never regain their proper shape. They are principally suitable as a passenger mask, in my opinion, though plenty of pilots use them with no penalty other than the possibility of losing some oxygen around the face seal, and perhaps a need for fairly frequent replacement.

A more durable and efficient type of mask is the Sierra, also sold as the AV-OXinc 249. It is made of a stiffer silicon rubber than the Scott mask and relies on a broad strip along the edge to effect a seal and cushion the mask on the face. It is held on by a broad web around the back of the head and a second strap over the top of the head which prevents the main strap from dropping. It is available with a built-in microphone, which is a considerable convenience, although as an alternative one can insert the boom mike of a Plantronics headset under the flexible edge of the Scott mask. Not very surprisingly, the AV-OXinc mask costs $70 with its oxygen connector ($110 with microphone), while the Scott mask, which should cost about $.50, in my opinion, costs $30 with connector (the connector alone in each case is $10.50). I think the AV-OXinc mask is the superior piece of equipment and the Scott mask is overpriced.

An alternative to a face mask is a piece of medical equipment called a "nasal cannula," which consists of two little tubes which one sticks up into one's nostrils. I have never tried one out; it looks uncomfortable, but perhaps it is no worse than a mask, only more intimate. The drawback of the nasal cannula is that it is a low-altitude, low-efficiency system which is "approved" only for short periods at up to 12,500 feet—in fact, it ceases to be approved at exactly the point where, according to the FAA, oxygen begins to be required. Its big selling point is that it is light, and it doesn't get in the way of a microphone.

Although turbocharged light aircraft often have built-in oxygen systems, nonturbocharged ones almost never do, and oxygen must be carried in the form of a "portable" bottle (I put the word in quotes because the bottles are extremely cumbersome) which is usually hung in a kind of sling behind one of the front seats.

Portable bottles of up to 50 cubic feet can be bought, with regulators, for between $350 and $450 new; 50 cubic feet should last one person ten hours, supposedly, though it is important to bear in mind that the duration depends on the altitude, assuming that the regulator is of the altitude-compensating type. It is important to give an oxygen system, like a fuel system, a practical test. One often hears the complaint that the oxygen bottle doesn't last as long as it is supposed to. Whether this is due to leaks, to incomplete filling, or to misunderstandings about advertised claims, it is good for a pilot to know what he can really expect from his system.

Besides oxygen, long flights call for other items of personal equipment. Earplugs are very desirable. Not only is the noise in airplanes eventually damaging to one's hearing, but it is also thought to contribute to fatigue. Plain cotton is ineffective as a noise barrier, but various other kinds of ear plugs are available. Those consisting of wax mixed with cotton are messy to use, and can melt in a hot map compartment. A Swedish product (the trade name is Billesholm), consisting of finely spun glass fibers, is effective, but not generally available. Rubber earplugs are good, but give some people pain because of the slight pressure they exert on the tissue of the ear for a long period of time; and some people have such large ear canals that they can't get rubber earplugs to stay in. I have had good service from the "Com-Fit" type of plug, consisting of a stem and three diaphragms of different sizes. However, I prefer for my own use the little cylinders of yellow closed-cell foam (E-A-R plugs) which one can use a number of times and throw away when they get dirty. They are both comfortable and effective. A couple of pairs of the rubber Com-Fit plugs, kept clean, are useful for passengers, or a supply of spare E-A-R plugs can be carried.

I have not found the type of silicon plug that is molded to the user's ear very satisfactory. I believe that the small hairs of the ear canal tend to hold open an air passage around the plugs, admitting noise.

One can generally contrive not to have to defecate during flights, however long, but sometimes it seems desirable to urinate. Men have no problem; any bottle will do. There is a product for women called "Jill's John," about whose practicality I have no information. A doctor once recommended to me, for the use of my female companion, an "orthopedic bedpan." I never checked that out either but it might be worth a try. Many women seem to experience urinary urgencies more frequently than most men, and their needs can be a serious consideration in planning very long flights. There might be some justification for segregating the sexes in front and back seats if the need to urinate en route is liable to arise; otherwise there can be problems about inhibitions, coyness, and bad jokes.

Bottles used to store urine should close securely. Attempts to dispose of urine overboard, for instance through the "storm window," are not advisable. If the bottle is embarrassing, a paper bag can be brought along to conceal it when you disembark, as a wino conceals his bottle.

The monotony of long flights is relieved if food is brought along. Needless to say, this is not the time to serve up vast quantities of prunes. Sandwiches, crackers, cheese, and some fruits are convenient. A thermos of water or fruit juice is a good idea, but it should be drunk slowly. The body transpires water steadily at a high altitude, and replenishment is desirable; but if you replenish too rapidly, it will just make you want to urinate. Carbonated drinks should have screw tops, not caps, because they will tend to foam up when opened at altitude, and the foaming will be more easily controlled with a screw top. Hot drinks, such as coffee in a thermos, are potentially dangerous; the boiling point of water is quite a bit lower at high altitude than at sea level, and a thermos of hot coffee, opened at 10,000 feet, may boil over, not only making a mess but scalding people as well.

On ocean flights where traffic is not a consideration, one can read and sleep if there is an autopilot or a copilot to hold a heading. Since it may be necessary to read for hours on end, a book is better than some magazines, especially because the idiotic content of most magazines does not accord well with the cosmic feelings one is liable to encounter in mid-ocean. I have had a lot of pleasure from a car stereo which I installed in my plane. It plays AM, FM, and cassettes through headphones, and it is sometimes a great help in passing the time. Every mention of FM receivers in airplanes, however, must be accompanied with the warning that they can interfere with VOR receivers, or so I am told. I have never seen it happen. If interference occurs, all you have to do is turn one or the other off.

A few small items: soft pillows to adjust the feel of the seat; a camera; a notebook. The late Max Conrad used to take a guitar and compose songs. Some people would probably suggest a Bible; but since so much time will be spent sitting, a book on Zen might be better.

Seat comfort can become a major problem on a very long flight. Trouble begins after three or four hours, if it begins at all. I have found it helpful to remove my wallet from my back pocket, and sometimes even to take off my belt, which tends, when I am sitting, to form a small, irritating kink at the base of my back. I avoid wearing blue jeans on long flights, because they are too tight and stiff. Soft, loose-fitting pants are better; skirts would be best for both sexes were it not so embarrassing for a man to disembark from an airplane clad in a skirt. Shoes that slip on and off easily are nice; I like to slip them off en route, and then I find myself trying to wrestle them back on over the outer marker.

From the force of habit I always keep my seat belt loosely secured around my lap, but I remove my shoulder harness once at altitude. Actually, the only reason for having the seat belt on during the cruising portion of the flight is to keep you from hitting the ceiling if you run into severe turbulence; but since the belt has to be cinched tight to hold you down, having it loosely slung over your lap really won't do you any good. Perhaps you could tighten the belt a little more quickly if you suddenly ran into some freakish weather.

One aspect of seat comfort which has received a great deal of lip service lately, but which was widely ignored until a few years ago, is lumbar support. The lumbar portion of the back is the incurving area from the kidneys to just above the buttocks. Most older airplanes have an essentially flat seat-back cushion, and as you become increasingly uncomfortable, your inclination is to slide forward in the seat, tucking the pelvis under the belly. This slouching posture provides temporary relief, followed by increasing pain. What the back is longing for, according to the present wisdom in seat design, is a thin cushion, an inch thick and six or seven inches high, in the small of the back, to help hold the spine in a position more similar to that which is assumed when you are standing or lying down. Small, lens-shaped cushions for this purpose can be purchased; they are intended for use in cars with poor lumbar support (which is to say, most cars). However, a soft pillow (stuffed with feathers, not foam) can work just as well, if you know where to put it. One does not intuitively appreciate the importance of lumbar support, and it is not even immediately apparent, when one inserts a support behind the lumbar, that it is doing any good. The difference is only obvious at the end of the flight.

Another important requirement for a comfortable seat is thigh support. The seat pan should extend to within a couple of inches of the knees; it should not stop halfway along the thighs. If the seat pan angle is adjustable, it should be sufficiently raised to put a light pressure along the underside of the thigh when you are sitting in your normal position. You should not sit on the points of the haunches, with the thighs elevated above the seat, as tall people are sometimes compelled to do. If the seat does not adjust far enough backward, or the pan does not tilt enough to give thigh support, then a pillow can be used here too.

Varying the seat position, back tilt, pan angle, and seat height from time to time will help reduce fatigue and discomfort. Isometric exercise is also useful; without moving the body, tense different sets of muscles against one another or against parts of the cabin, and hold the tension for fifteen or thirty seconds. A routine of exercises, repeated every hour, not only refreshes the body, but kills a little time as well.

If an airplane's seats do not seem very comfortable, it may be possible to replace them with seats from a later model of the same airplane or perhaps, with a little work, with seats from some other airplane. Volvo and Alfa Romeo sedans are said to have excellent seats, and one company, Recaro, sells luxurious replacement seats for cars; but they are too bulky, and certainly too heavy, for installation in an airplane. A geologic survey pilot of my acquaintance who flies ten hours a day in a Navajo has high praise for that airplane's seats, which he says are common to all Piper twins. Before planning to reshuffle seats, however, you should realize that to adapt a seat from one airplane to another of a different model or type may be a very complicated job which few shops will care to undertake.

Sheepskin covers are said to do miraculous things for seats, keeping your backside warm in winter, cool in summer, free of perspiration, and gently massaged to boot. I have sheepskins on my seats, and while I find them perfectly satisfactory, I do not find them miraculous; but it has been so long since I have sat for hours on a leatherette or fabric seat in an airplane that I cannot say for certain that the sheepskin is any better or worse. It is hard to get it cleaned; the hide seems to want to disintegrate along the seams.

No doubt the more comfortable you are the drowsier you may become; fighting off sleep can become the overriding concern on a long flight. Because of the risks of fatigue, it is always desirable to start a long flight early in the day, unless it is so long a flight that it would continue on into the wee hours. In that case, I would be inclined to start in the afternoon and arrive in daylight. Daylight may not guarantee that you'll stay awake, but it at least makes the approach and landing easier.

Although it is highly forbidden to suggest the use of drugs while flying, very long flights do put exceptional strains on pilots, and on prohibitions as well. Benzedrine was given to pilots during World War II. I am told that it was common practice among Navy pilots during the Vietnam War, when they flew nine-hour missions with three in-flight refuelings, to take an upper before returning to their carriers. Amphetamines are sometimes used by long-distance truckers, and while it may be that an accident is occasionally caused by some kind of drug-induced blunder, more accidents probably result from running out of drugs and falling asleep at the wheel. The same would be true in flying; it may be risky to adulterate your bodily essences with drugs; but it is probably equally so to fly for twenty hours and then make an instrument approach with your eyelids propped open with paper clips. I have never used an upper myself while flying—though I have not hesitated to drink a couple of cups of coffee during a rest stop—and I can't suggest any brand names, but at any rate I think it worth knowing that an upper will give a temporary boost to your alertness and abilities. The effect is only temporary, and when it ends, it usually ends with a dull thud, so the idea would be to land before the effect of the drug wore off. I would imagine—speaking as a layman reluctant to broach the subject to a flight surgeon—that the hazard to a pilot would not be in taking one upper at the end of a long flight, but in taking one every three hours for the better part of a day, or in having one wear off just at the end of the flight.

Most long flights can be completed in ten or fifteen hours, because most light aircraft, even equipped with cabin tanks, can't carry enough fuel to go longer than that. If you take off fairly early in the day, such a flight does not subject you to extremes of fatigue. Occasionally, however, an extremely long flight is attempted, for instance in pursuit of a record, and then fatigue becomes a more serious consideration than almost any other.

An instructive example of what I mean is provided by Dick Rutan, an ex-Air Force fighter pilot who set a closed-course distance record of 4,800 miles in the homebuilt canard airplane called the *Long-EZ* which his brother, Burt, designed. Both brothers took great pains in planning the flight, ensuring the reliability of all the systems on the airplane, and determining speeds and fuel flows with great precision. Everything that could be planned was planned. But there were imponderable elements which turned out to play a very important role.

Rutan took off a little after dawn. The course he flew, back and forth between Mojave and Bishop, California, paralleled a steep mountain range; in fact, he had to make a slight dogleg to avoid Mount Whitney, the tallest mountain in the contiguous United States. The terrain below was desert, almost devoid of lights and landmarks visible at night. He had no autopilot. By the middle of the night, Rutan was beginning to struggle against sleep; he could feel himself slipping down its smooth slope, and again and again he dragged himself back upward, only to slide back. He began to hear organ music and voices in the engine roar. The dark mountain masses around him resolved themselves into phantasmagoric threats. A spaceship pulled into formation off his right wing several times; when he turned to look at it, it scooted away. A small, ferocious man seated himself on the canard and began suggesting heading changes.

To the wakeful reader reflecting upon these weird events, their very unreality seems like sufficient protection against them. How could a pilot follow the advice of an obvious hallucination? But the hallucinator sees things differently. When the gremlin suggested to Rutan that he dive toward a nearby mountain, Rutan began to reflect that it was possible that he was already dead, having collided with the ground in his sleep. Therefore, this little fellow was part of the afterlife, and there could be no further danger in heeding his advice. Incredibly, Rutan considered making the suicidal turn on the grounds that he was already dead.

Preoccupied with his strange experiences for a long while after the flight, Rutan discussed them with a psychologist who attributed them to sensory deprivation. The lack of input from the peripheral nervous system gradually sends the brain into a narcotized state in which it fails to distinguish between reality and fantasy. Hallucinations become continuous. It is not necessary to be unhealthy or even to be severely deprived of sleep for this to happen; sensory deprivation, fatigue, boredom, and low-grade hypoxia (arising both from altitude and from immobility) have a synergistic effect.

For extremely long flights—Rutan's lasted until the following evening, over thirty-three hours—planning must include the mental condition of the pilot. The hours of darkness are dangerous. Trivial tasks—radio communications, navigational chores, calculations, course plotting—are valuable if they keep your mind functioning. Periodic doses of oxygen, even at altitudes below 10,000 feet, and especially at night, are valuable for two reasons; first, they may increase alertness and visual acuity; second, they may make you aware of hypoxic symptoms which have already crept over you unawares. Singing loudly and uttering bloodcurdling screams (most suitable for the solo traveler) are also helpful for staying awake.

Rutan, incidentally, had a couple of tablets of a stimulant drug aboard, but he forgot to use them. Considering his experiences during the flight, it is obvious why the hazards of taking a drug while flying are minor compared with those of remaining awake in a droning airplane for very long periods. It is also obvious why an autopilot might be extremely useful. A series of short sleeps—you can set one or two alarm clocks to make sure you wake up promptly—can do a great deal to alleviate fatigue. Obviously, there is some small possibility of combined autopilot and alarm clock failure; but one must weigh one set of dangers against the other, and it seems to me that the chance of serious trouble from extreme fatigue is much the greater.

3

OVER-OCEAN FLYING

Professional ferry pilots are nearly unanimous in discouraging inexperienced pilots from attempting to fly across oceans—namely the Atlantic and Pacific, since trips to Europe and Hawaii are the ones which immediately occur to American pilots. But the desire to make such trips is there, and so many people have made them successfully, sometimes with very modest aircraft, that it is useless to behave as though such flying were folly. It has been going on for a long time, and will continue to go on; the successes far outnumber the failures, and most of the failures can be traced to avoidable miscalculations and misjudgments. The penalty of failure may be only the expense and inconvenience of search and rescue teams, usually military; it could be loss of an expensive airplane; or it could be death for all on board. The light airplane pilot who wants to take his family to Europe must be aware that he is playing for large stakes; the trip cannot be made casually. But it can be made, and made safely, on condition that it is taken seriously.

Experience is the single great element that separates the professional ferry pilot from the novice; it may not sound like a big thing—it is so imponderable and imprecise, and besides everyone was inexperienced at everything at one time or another—but it is experience that enables the pro to make with ease decisions over which the beginner must agonize, and that enables him to feel in his bones what must be discovered by the beginner through reason and calculation. People forget what it felt like to be inexperienced at the things they now take for granted; they forget the anxiety and doubt of inexperience, and all the stupid errors a beginner can make. Pilots who have flown a few hundred hours have already lost touch with the first solo cross-country flight, with its "lost in the sky" feeling and its host of imaginary dangers. But the over-ocean flight can be as bad and worse, and with much more at stake. The novice fails entirely to consider, for instance, while at home contemplating a globe—sweet fruit to a dreamer's palate—that ten hours out of San Francisco, ten short of Honolulu, he could suddenly be seized with a panic fear that would set his hands shaking, his teeth chattering, and make him unable to read a compass or understand a map, all because of suddenly grasping, in some way that he hadn't grasped it before, that he is all alone out here, incredibly remote from fellowship and help, with a single engine on which everything depends, and only the floor of the ocean, creeping with crabs and vermin, on which to rest.

It is necessary to think about an ocean flight for a long time ahead. Ferry pilots go back and forth, back and forth, and prepare each flight as it comes, just as one does over land. But a beginner has no school but patient study.

It is necessary to talk with people who have done it; look up old magazine articles; read about navigation and meteorology in the encyclopedias and textbooks of the local library; write away for brochures; shop for equipment; study the globe and the charts until you could navigate without them; in other words, exhaust the subject. You have to become conversant with details, and this is done simply by absorbing everything you can; experience will later allow you to sift out the essentials and discard what is superfluous. But what is important is that once you actually set out, the way not be strewn with surprises. For a pilot who has never attempted an ocean flight before, a period of several months of mental preparation is not excessive.

The first step is to decide at what time of year you will make your trip. Two tactical considerations come into play: wind and weather. Since your studies of weather will be in the general statistical realm—you will be studying "tendencies"—you must always bear in mind that the "facts" you learn will be merely statistical in nature, and the weather you actually encounter may be quite different from the norms found in books. But you have to plan on the norms. You must, for example, assume that the North Atlantic in winter will be full of icing and storms, even though you hear that

The real challenge of long-distance flight—a transoceanic voyage.

ferry pilots continue to cross it during the winter, fitting their flights between fronts. You must never count on good luck; you must, on the contrary, plan for bad.

A good source of basic information is a large world atlas. Some of them include lat/long listings for cities in the index, as well as seasonal maps of prevailing wind flows, precipitation, temperatures, and so on. A city library should offer a variety of atlases for reference and give you a chance to see what is available; you may then buy a suitable one at a bookstore—possibly a used bookstore since many of the modern mass-market atlases lack the detailed information, like lat/long listings, that is of special interest to a pilot. The New York Times Atlas, for instance, is full of information about the evolution of planets and the world automobile industry, but it contains a minimum of meteorological intelligence, and virtually nothing on the subject of winds.

Textbooks on basic meteorology and climatology always contain sections on world winds, seasonal variations, important local weather patterns, the locations and movements of so-called semipermanent highs and lows, storm tracks, and so on. It is generally useful to know that in certain latitudes the flow is in one direction, and in others in another; that there is an Aleutian low, a Greenland low, a Pacific high, and so on. The weather one actually encounters, of course, need bear no resemblance to the general weather pattern.

Next, there are sources of detailed meteorological information about specific stations. They usually contain a digest of information collected over many years, and they present it as a monthly breakdown. Some can be found in libraries; records for specific stations can be obtained from the U.S. Weather Service.

In reviewing my own records of my preparations for various trips, I found the following page which I reproduce to give the flavor of the intelligence derived from a visit to the library:

> Aleutian weather.
>
> Visited LA Pub Library on January 15. It appears that the early summer months (June/July) offer the best combination of above-freezing temperatures and reasonable cloud cover and visibility for the eastern end of the Aleutian chain. Records for Cold Bay indicate almost continual overcast in June and July, but heavy fog (less than 1/4 mile visi) on 2–6 days a month. Generally the winds are crosstrack. Out at Shemya heavy fog is almost continual during the summer; Shemya would be suitable only as an emergency strip unless the weather were accidentally good when we passed through.
>
> There is comparatively low precip in most of Alaska during these months.
>
> The Aleutians lie precisely on the North Pacific storm track.

> From Cold Bay to Kushiro is about 2410 sm. As the weather is very different in the eastern Aleutians from the western, it may be possible to use an airport farther west than Cold Bay, such as Dutch Harbor, North Shore, or Umnak, as a jumping-off place; but Shemya seems practically out of the question. It is still necessary to find out about the northern Japanese weather during June/July.

While this brief note to myself is no doubt a piece of work inferior to what a professional meteorologist might produce, it nevertheless was valuable, and it is a good indication of what a couple of hours at a library will produce. Accompanying it were photocopies of weather digests for Adak and Shemya, and July and August charts of wind "isotachs"—lines of the same average velocity—and prevailing directions for the Pacific for the 850 and 700 millibar levels. From these it was apparent that the prevailing wind in the vicinity was westerly, and that although it reached its peak velocity at a location southwest of my proposed track, that location was not far removed from my track and so the chance of a higher than normal headwind was, I reasoned, great. Furthermore, the area was far from any of the so-called "singular points" which are centers of apparent convergence or divergence of winds, and which suggest a certain variability in the local flow. Such a point lies between California and Hawaii, and indeed the winds along that route tend to be variable, and to change direction at some point on the trip. The Aleutian winds, on the other hand, looked determined to flow one way, and not the other.

Flying weather charts for Adak and Shemya, which I found in the library, showed a high incidence of fog in the summer months, and nearly constant restrictions of ceiling and visibility; October was the best month in both locations; September was not too bad. Since in-flight icing was my main worry besides being delayed by below-minimum weather, I would have liked to make the flight in late August or early September; as it turned out, I made it in July, which was according to the charts the worst month for fog, though the best for temperature. In defiance of the literature, the weather on the particular day I flew to Japan was mostly clear, with a strong following wind due to a chance low-pressure center moving up from the central Pacific.

It seems to me that there are two lessons in this. One is that if you take the trouble to seek out weather information in a big city library you can find surprisingly detailed data; the other is that when you get there, the weather is what it is.

Track plotting for ocean legs can be done on maps of a very small scale; all you really need is room to note heading changes at convenient intervals. If you use the electronic calculator method of determining headings and distances, which I will describe in the next chapter, the accuracy of the chart is not important, and so neither is its scale. It is useful, however, to use something a little more accurate than a sketch of the Pacific Ocean on a cocktail napkin, because before taking off you should cross-check all calculated headings and distances against the chart with a protractor, looking for gross errors like incorrect application of a magnetic variation correction.

The choice of planning charts is small; they come from the Department of Commerce National Oceanic and Atmospheric Administration, National Ocean Survey, Riverdale, Maryland 20840. You can write for a catalog of the charts available, which contains information on the scale, coverage, and type of charts, as well as ordering information. For general planning within the United States, there are VFR/IFR planning charts (VFR on one side, IFR on the other), which are quite cumbersome and are meant to be hung on a wall. The charts come in two pieces; when joined, they measure 56 by 82 inches, and make the country look harder to fly across than it really is. I prefer to rough out my route on the Jeppesen Low Altitude Planning Chart, which presents a network of principal airways and gives an instant visual impression of the most convenient way to break a long trip into segments, and of the airways to follow. This procedure has certain shortcomings, however; for one, it gives you the impression that the only places to stop are big cities and omni stations.

For Caribbean flights, there is a planning chart of the same format as the one for the United States, but on a slightly larger scale.

The VFR sides of these charts show relief, in case it is of any use. For some airplanes, the presence of a 14,000-foot mountain barrier is a serious consideration; for others, not. Since it may save a little time and fuel to avoid making unnecessary climbs and descents, the VFR relief information can be useful in selecting the highest elevation at which to make a refueling stop (density altitude permitting). At any rate, the VFR charts are very slightly more decorative than the IFR, though they are not in the living color of the sectional and WAC (World Aeronautical Charts) charts.

For planning and inspiration for international flights, the GNC (Global Navigation Chart) series is nice; it has colored relief and basic geographical information (some cities, airports, radio beacons, roads, and so on) on a scale of 1 to 5,000,000. The North Pacific and North Atlantic are covered by GNC-6 and GNC-3 respectively; GNC-8 covers half the United States and the Pacific out to Hawaii; GNC-9 includes most of the United States, all of the Caribbean and Central America, and the northernmost portions of South America. Some routes are not so conveniently covered; the South Atlantic route between Brazil and western Africa, for instance, occupies the edges of two different charts, which furthermore have almost nothing on them but water. You could do nearly as well with some blue wrapping paper.

The normal procedure is to draw lines representing your proposed route on a GNC, tack it to the wall, and stare at it for hours.

For VFR navigation by pilotage, there are ONC/WAC charts. The ONC series (Operational Navigation Charts) are slightly prettier than WAC charts (they have "shaded relief"—the fanciful shadows of mountains—in addition to contour and height color coding), but otherwise the same, and except for the United States and portions of western Canada and eastern Siberia, all the land areas in the world are covered by ONC charts.

Unless you want to carry a file cabinet with you in your airplane, it is pointless to take along entire charts of portions of your route which contain only one or two islands on which you don't intend to land anyway. If you were flying from Alsaka to Japan, for instance, ONC E-11, E1-10, and F-11, which show portions of the Kamchatka Peninsula, Kurile Islands, and the extreme western Aleutians, would be largely dead weight. However, you can have the information on them without carrying the entire chart around by making up a strip chart covering your route and the land areas within a hundred miles to either side of it. You cut out the portions of the charts which may be of use, tape them together, and accordion-fold the strip so that you can move along it while keeping it folded to a convenient size. This way you have VFR information which may be useful if you experience a radio failure, stray from your course and sight an island, or, for that matter, detour for some reason to make an unplanned landing.

In making strip charts, it is good to circle all airports and to note on each segment of the chart, as folded, the heading of the intended route and the progress along the route in 100-mile increments. Thus it is possible to see, without unfolding the map, where you are, what is in the vicinity, and which way you should head in an emergency. All this is perhaps rather childish busywork, and much of it is bound to be useless; but this kind of very thorough planning is an enjoyable part of a long flight, and it is also instructive.

For position plotting over oceans, the chart usually recommended is the Aircraft Position Chart: number 3097 for the extreme North Atlantic (via Greenland and Iceland); 3071 for the middle North Atlantic, including Gander–Shannon and Gander–Azores; 3094 for the North Pacific; 3096 for California–Hawaii; and so on. Indispensable for Loran and Consol navigation, the Position Charts are otherwise rather unsatisfactory, consisting only of a maze of colored lines of position with the ghost of a terrain map underneath. More compact and equally useful for keeping track of position are the Orientation Charts from Jeppesen. They show Consol information, frequencies and identifiers of radio beacons, and basic geographical information, as do the Position Charts, but they include HF frequencies as well. More important, they are more compact and easier to read than Position Charts are.

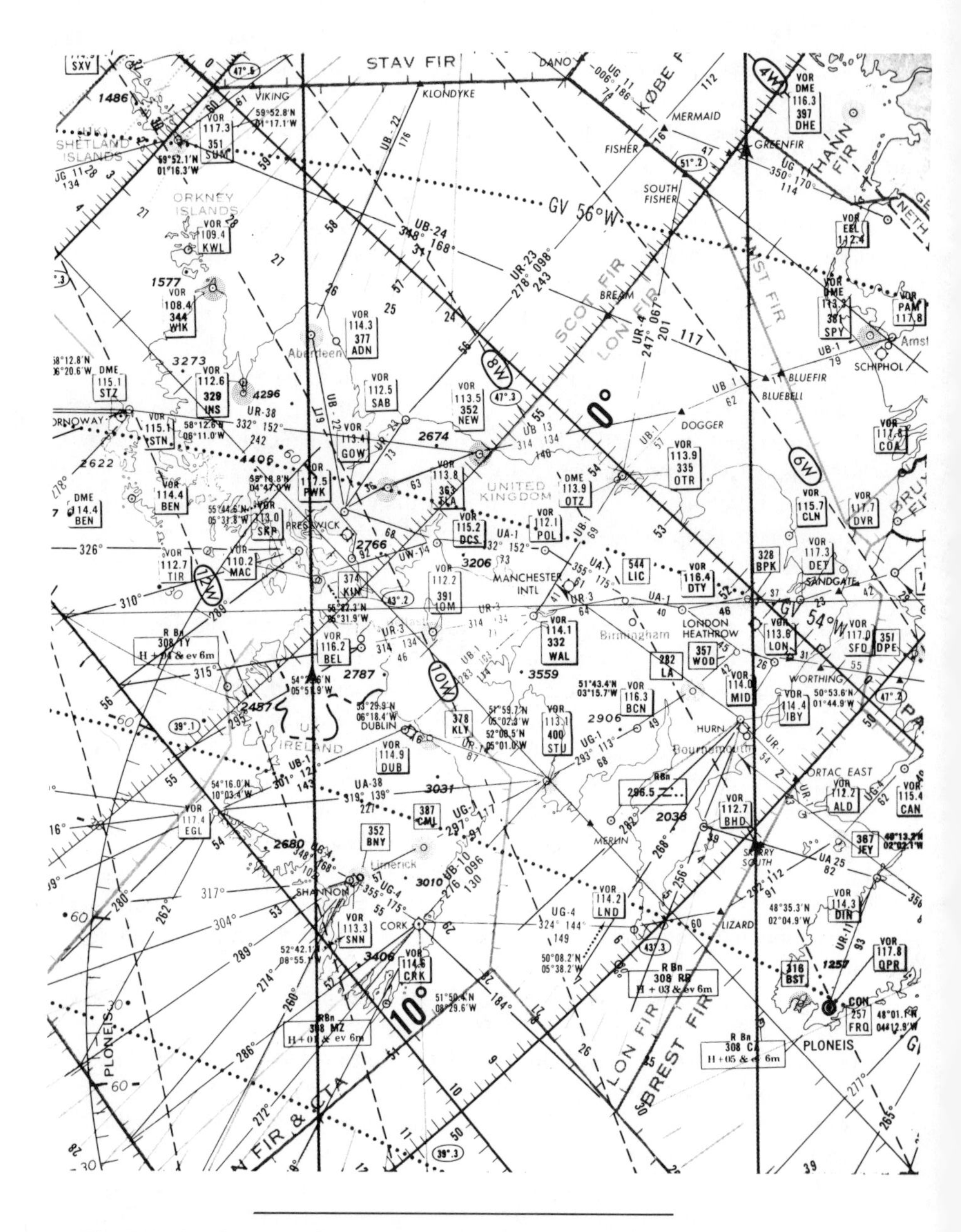

The British Isles according to Aircraft Position Chart 3097. Land areas are barely discernible under all the Loran information.

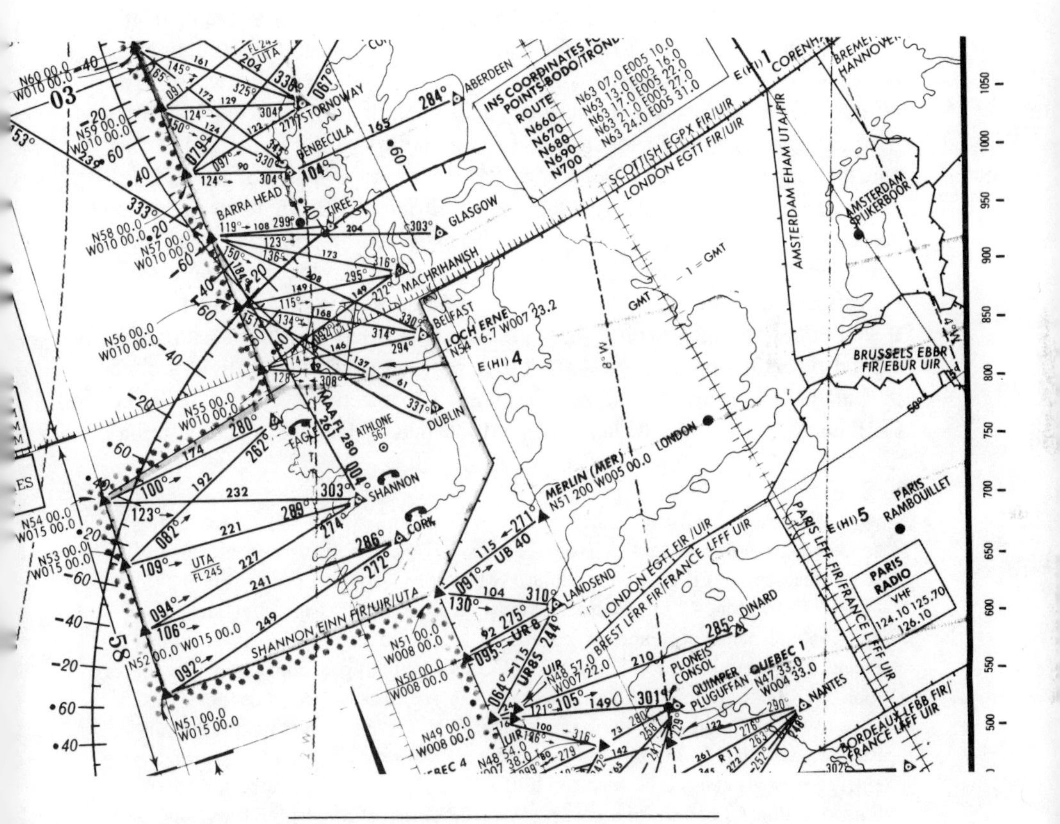

Jeppesen's version of the same area, with everything you need to know unless you are using Loran.

In my note to myself about the Aleutian weather, I speculated about the possibility of using several other airports as jumping-off places, rather than Cold Bay. I later learned that Cold Bay was the only one with a hard-surfaced runway of sufficient length for my purposes. Runway length is shown on the ONC charts, but more complete airport information is found in the Department of Defense FLIP (Flight Information Publication) Supplements, which give a vast amount of data about airports all over the world. The information is tailored for military use and is a little difficult to figure out at first; but certain basic facts, such as the availability of the right kind of fuel and the airport's hours of operation, are there. Since many of the airports along remote ocean routes operate rather differently from ones in busier parts of the world, this information can be extremely important. Let me give one example. A frequently used landing place for airplanes en route from Hawaii to Australia or other such points is Majuro, in the Marshall Islands. If you talked casually with a ferry pilot he would tell you that Majuro was a good place to stop, and you might then proceed to plan your flight around it. But there is more to landing at Majuro than meets the eye. Here is the FLIP entry for Majuro, which is officially called "Marshall Islands International":

> . . . Tran acft must make prior arng for fuel with Mobil Oil Guam. After sender has confirmed fuel delivery he must give 24 hr. advanced ntc to arpt superintendent and IMG officer Majuro, Marshall I. If ETA of ldg is btn 0400Z Fri to 2200Z Mon, 48 hr advance ntc must be given to arpt superintendent. Msg will include name of sender, type of acft . . . and that sender has obtained fuel delivery confirmation fr Mobil Oil Guam including quantity and type of fuel. . . . Due to pwr limitation all ngt ldg are discouraged. Ldg perm will have to be apv in advance by the arpt superintendent and/or by the Dist Administrator, Marshall I. Emerg will be accepted as needed, but the nature of the emergency will need to be reported to the Director of Transportation and Com, Saipan within 48 hr. after its occurrence. Authorities at Majuro will investigate the validity of the emerg and either hold the acft or clear its release. . . .

There is more to a stop at Majuro than you might have thought.

The note about emergencies is interesting. Many airports along ocean routes and in remote places are military fields or are under some kind of military administration. They require prior permission for landing, and in some cases refuse all civil aircraft, or all which are not on "bona fide ferry flights" or record attempts. However, any airplane in an emergency is automatically given permission to land; and so quite a few pilots have declared emergencies of one sort or another in order to secure permission to land, figuring (correctly) that they would talk their way out of the situation once they were in it. This proceeding has become quite familiar to the military authorities; hence the admonitory note regarding Majuro, which applies equally to a lot of other places. I find the reluctance to accommodate civil pilots puzzling—surely very few will ever come their way—but it is a tradition of long standing.

The exception for record attempts might be useful in some cases. For several hundred dollars in fees you can turn a trip into an official record attempt (see Chapter 9) and thereby qualify for special consideration by the military authorities. You could think of the cost as a huge landing fee. Besides, records are there for the taking, and it would not be difficult to set up your route to include a few.

In addition to the elaborate requirements for landing at certain remote airports, there are requirements and regulations for flying abroad and over water which call for considerable study. The demands of each national government are stated in the *International Flight Information Manual* (the *IFIM,* obtainable from the Department of Commerce at the same Riverdale, Maryland, address given for the charts discussed above). Permission for overflight and landing is required in Latin America and for many countries in the Eastern Hemisphere, those of Europe, down as far as Turkey, excluded. In general, however, an aircraft with a normal certificate of airworthiness can travel about the nonsocialist countries of the world more or less at will, subject to import-export requirements. So far as the pilot is concerned, a U.S. license is sufficient, unless he intends to fly foreign-registered aircraft while abroad.

Homebuilt or other experimental-category aircraft require letters of authorization from foreign governments. These are not difficult to obtain, so long as complete information regarding the aircraft, including copies of all its official documents and those of the pilot, is sent along with the request. The only problem is to make it clear to the foreign administrator what exactly you are asking for, since the permission is an FAA requirement which has never occurred to some foreign officials. For instance, Mexico does not require special dispensation for amateur-built aircraft on tourist trips. A possible source of trouble is the long delay that often intervenes between the request for permission and the receipt of it. By sending requests by certified mail some sense of urgency may be conveyed to the recipient. Wires are sometimes successful, sometimes not at all. Two months is a reasonable delay to expect.

For single-engine flights departing or overflying Canada for Europe there is a list of Canadian Government requirements, compliance with which must be demonstrated at Moncton on the way to Gander or Goose. The requirements are in some cases a little fussy (e.g., either two landing lights or two filaments in one light) but they are intelligently applied by congenial officials whose sense of their mission is to ensure the safety of travelers, not to enforce the letter of the law. At Moncton you are subjected to a short oral test on your knowledge of oceanic navigation, search and rescue procedures, position reporting, and so on, and your equipment is inspected. If you fail the test, you receive further instruction on the spot. The test is used only as a way of finding areas in which you are weak, and filling you in on them.

You have to have your papers in order at Moncton, including a ferry permit if the airplane is operating at a weight in excess of its normal gross; instrument rating; and FAA Form 337 for additional tanks. Airplane equipment required includes a heated pitot tube, a compass that has been swung within the past thirty days, blind-flying instrumentation, HF radio, charts, and two low frequency/medium frequency radios. Fortunately, this last requirement is interpreted to include one hand-held broad-band unit; the other unit has to be an installed ADF. The list of survival equipment is long but well thought out, and includes life jackets, portable and buoyant ELT (Emergency Locator Transmitter), flare gun, raft, and a set of items that should be secured to the raft. This list is a good basic one, and all commercial raft systems include all the items in the list.

The HF radio requirement is a troublesome one for a person planning to make only one oceanic crossing. HFs can be rented, but rental is paid by the day, normally, and if you plan to spend any time abroad you would end up paying a great deal for a day's use of the radio. Sometimes you can locate a ferry pilot of the company from which you rent the radio, and send it back with him. Also, you usually have to leave a deposit of the replacement value of the radio—$1,200, the time I rented one. I might add that the one I rented didn't work most of the time. HF makes use of a long wave which is readily channeled by the layers of the atmosphere and can travel distances of many hundreds of miles—well beyond the line of sight. A long wave requires a long antenna. There are two approaches to the HF antenna problem. The more usual is to string a wire from wing tip to fin to wing tip, with a pigtail coming off one run and going to the cabin, where it runs into a box called an "antenna coupler," and thence to the radio. The coupler is used to tune the radio to the antenna installation. An alternative arrangement, mechanically less convenient but electronically superior, is the trailing antenna, which can be reeled in and out, and its length set to the precise value for the best signal at the frequency being used. The length of the antenna is set by watching for a peak reading on a meter on the radio face. On a temporary installation, the mast from which the antenna will be deployed protrudes from a hole drilled in the cabin floor in front of the pilot's seat. The pilot reaches down and winds the wire in and out on a reel; a small funnel attached to the end of the wire provides the drag to make it trail behind the airplane. The radio tuner is then put in some convenient place in the cockpit—if there is a convenient place.

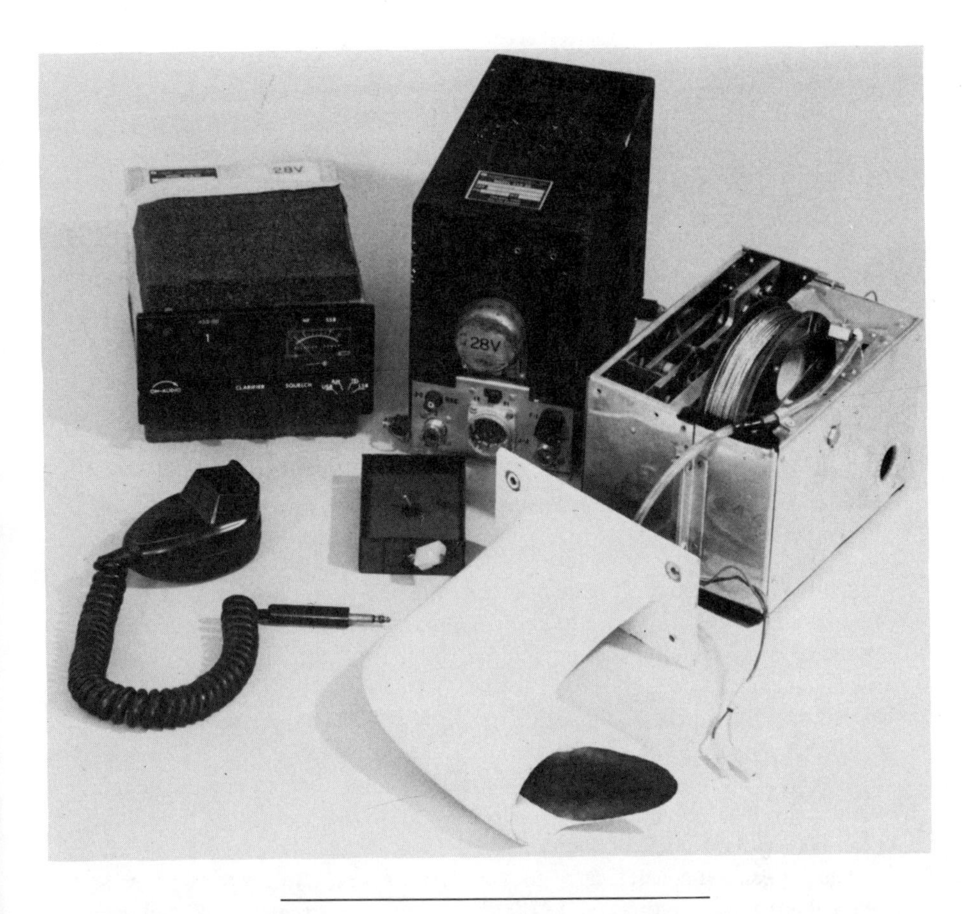

A typical HF radio, this one the author's Sunair ASB-60. The antenna reel and mast are homemade and do not follow the normal lightplane practice of stringing an antenna from wingtips to tail.

HF operates in two modes, AM and SSB. SSB, or single side band, is clearer and more powerful. Not all stations have SSB capability, but it is very desirable for an HF radio to be able to work in either mode.

HF is crystal tuned, and there is usually a separate crystal for each frequency to be used. Different areas of the world have unique groups of assigned frequencies; since you can't store enough frequencies for the whole world in any but the most expensive sets, you select a couple of frequencies for each area in which you anticipate flying—or perhaps six for one area, if that's the only place you'll go. Although the frequency band runs from 2,000 to 22,000 kHz (kiloHertz), the most successful frequencies, according to my sources, are around 5,000 to 9,000 kHz; and so I have tried to pick one from each end of that band in each area that I expect to use. But I own my HF set now; if you rent one, you simply rent it in the area of use, and it will automatically contain the necessary frequencies. You merely select channel 1, 2, 3, and so on, according to a list that is included with the radio.

If you are using a reel antenna, the required length is inversely proportional to the frequency. It is helpful before the flight to measure the circumference of the reel, in order to know approximately how many turns correspond to what length of trailing antenna. In general, the length of a quarter-wave antenna is roughly equal to 250,000 divided by the frequency in kiloHertz. 5,610 kHz, for instance, requires an antenna length of 250,000/5,610, or about 45 feet. Some efficiency can be gained at high frequencies by using a three-quarter wave antenna length, or three times the length yielded by the formula. Thus, for 13,336 kHz, the best length would be (250,000/13,336) × 3, or about 56 feet.

The higher frequencies carry for longer distances, and range is better at night than in the daytime. A 17,000 kHz signal may range 3,000 miles at night, a 3,000 kHz signal only 150 miles; but on the other hand while the 17,000 kHz signal may be audible 3,000 miles away, it may not be audible 300 miles away. Still, an 8,000 kHz signal may possibly also range 3,000 miles at night; and at any rate there is not much point in having a nominal broadcasting range much greater than half the cruising range of the airplane. It is for this reason that the frequencies around the middle of the HF band are the most versatile ones.

HF sets are rather heavy—twenty to thirty pounds, all parts considered. When I bought one, I installed the wire harness and racks for it in my airplane, but arranged everything so that the radio boxes, reel, and antenna could be quickly removed. I used the trailing type of antenna for two reasons; one was that my airplane was too small to allow an optimum fixed antenna; the other was that with the trailing antenna, the coupler box, which weighs quite a bit and takes up a lot of space, could be eliminated. I also

HF coverage for the North Atlantic. Depending on your route, you would select frequencies from among several available for each zone (Nat A, Nat B & C, etc.), and then have crystals installed in your radio for each one. Most of the less expensive HF units have only six or ten channels.

frankly enjoyed the challenge of building the remote-controlled reel; though when I look at it now, I am astonished that I ever did it. The reel, mast, and antenna weigh together nearly six pounds; so the weight saving over a coupler box was not great. The mechanically simpler fixed antenna is in wide use, and is much preferable.

HF sets are expensive; the least expensive ones are not very good, and the good ones are priced at upward of $3,000 new. Very few are in use in the contiguous United States, and the chance of finding a used one to buy is slim. On the other hand, many are used in Alaska, and a search of ads in *Trade-a-Plane* might turn one up. A good unit will cost between $1,000 and $2,000 used. Remember that SSB capability is very desirable, and that 100 watts of SSB power is the equivalent of several hundred watts of AM power under poor conditions.

While I had difficulty in ever raising anybody on either side of the Atlantic with my rented HF, I easily spoke to Anchorage from just north of Japan with the one I installed in my airplane. No doubt my installed unit is a better one in many ways, but there is also a certain element of chance in all HF communication. Sometimes the thing just won't work; there is too much static, or it makes noises like a school of fish, or it cuts in and out, or they hear you and you don't hear them or vice versa, or nobody hears anybody. In that case, you fall back on VHF relays through airliners, which you contact on 121.5, switching to 123.45 once you have established contact. (All aircraft are supposed to guard 121.5 continuously over the ocean, so you can often raise someone with a blind call, but you then should get off the emergency frequency to conduct your conversation.) Sometimes an airliner will be able to make HF contact when you cannot; sometimes it's the other way around. In any case, they are usually happy to do you a favor, and eager to express amazement at your folly in being out there.

You may wonder why, if there are airplanes to relay through, you need HF in the first place. The answer is that, first, you can't always make a relay; and second, the rules require that airplanes be equipped to do their own communicating, all ocean-crossing flight plans being IFR. The exception is leaving the British Isles; Shanwick (Shannon-Prestwick, the oceanic control authority for that side of the ocean) doesn't mind if you file VFR (below 5,500 feet) westbound out of Shannon. You can even give VFR position reports by relay, if you like; but you must remember always to say you are below 5,500 feet. Conversely, once you are out in the middle of the ocean, it is difficult to remember that you are on an IFR clearance, but the controllers want to give you permission to make altitude changes. In order to keep them happy, ask for their say-so before changing altitudes. If you want to drop below 5,500 feet, however, you will be leaving controlled airspace and cancelling IFR.

Sometimes, especially if icing is a problem, it is desirable to fly down low. In that case it is useful to know that flight altitude does not affect the quality of HF communication—so long as you don't drop your trailing antenna into the water.

VHF communications over water are often surprisingly good—better than line of sight—and you may pick up VOR signals or voice at a distance of 150 miles while flying below 10,000 feet. Because of the strength of VOR signals over water, it is a good idea to plan flights to leave a coast at a VOR, and to make the first landfall at one. When leaving, the VOR gives an opportunity to check the wind drift (you have to assume that your compass is accurate; it should have been swung, or at least checked, with the HF installed in the cockpit and operating) and makes it easy to hold track for the first half hour or hour of the flight. On arrival, the VOR makes it easy to correct your course to arrive through the proper gate.

ADF is indispensable on overwater flights. Most North Atlantic terminals do not have VOR, and in the Pacific the long range of NDBs is your best insurance against missing your destination. Your ADF should be checked for accuracy and sensitivity before a trip; accuracy in pointing is less important than sensitivity. I have heard a lot of good talk about the old Bendix ADF-12; some of the newer units do not have such a good reputation. One that does is the ARC (Cessna), which is sometimes mentioned as the only good radio in that company's line. I had a Narco ADF-140 which was all right, but nothing special, and required an external sense antenna; I replaced it with a Bendix ADF-2070, which uses only a single-blade antenna which it shares with one comm radio. I have been very satisfied with it. It has an "extended range" feature that uses some kind of electronic trickery to search through the noise for a coherent signal, and that seems to give it very good sensitivity.

One of the weaknesses of ADF is its susceptibility to coast effect—the bending of radio waves as they pass a shoreline. Since all waves coming to you from coasts and islands are going to have passed a shoreline, there may be errors in their indicated bearings. If you are tracking toward the beacon, these errors may take you a little out of your way, but they are not a serious problem. Tracking away from a beacon, they could be more significant. They can also affect time-distance calculations. In general, however, the use of ADF in ocean flying is quite basic; you want the needle to point ahead of you when you are going toward the station; and if, for instance, you are heading from Iceland into Goose, you want the needle to point to one side of the nose for Cartwright, to the other for Hopedale. That is enough.

AM broadcast stations are often far more powerful than aeronautical beacons; the problem is to know where they are. A book entitled *World Radio and TV Handbook* from Billboard Publications, 1515 Broadway, New York,

New York 10036 lists all the major broadcast stations in the world, giving their signal strength and location by town name. Sometimes it is not easy to match up a town name with a location on a map, but the choice of stations is quite large, and they are good for entertainment as well as for navigation. The BBC on 400 kHz is a powerful signal usually received in the mid-Atlantic, if not sooner, on eastbound flights.

Apart from survival equipment, which I will discuss elsewhere, the *sine qua non* of ocean flying is usually supplementary fuel. It is possible to cross several large bodies of water, including the North Atlantic, with the standard tankage of light airplanes, but only by making a series of short hops. Since overwater legs require unusually large reserves because of the lack of airports en route at which to land in case of headwinds or bad weather, the usual procedure for ferry pilots, and one which amateurs do well to adopt, is to supplement the tankage of the airplane with one or more tanks carried in the cabin.

Ferry flights usually go only one way, and ferry fuel tanks are normally made to use once and then discard. Some services use simple fifty-five-gallon drums in cabin-class aircraft; others have tanks made up of galvanized steel. Weight is not a serious consideration, nor is the efficient use of space. For the owner-pilot who wants to make a round trip across an ocean, there are several options. One is to obtain a ferry tank from a ferry service and keep it in the airplane for the entire trip. Another is to have a tank made specially, or to make one himself; this is not so difficult as it may sound, but it is hardly a project for someone who has never done any shopwork before. There are custom tank manufacturing firms in many cities. Tanks can be made of fiber glass, aluminum, or steel; the advantage of a custom-made tank is that it can be sized to make the best use of cabin space, and if you contemplate making more than one long overwater trip, a fitted system designed for indefinite life is a good idea.

Single-engine airplanes are easier to fit with extra tanks than small twins, because they get better gas mileage, and their cabins are usually little smaller than those of light twins. Remember that any significant amount of additional fuel has to be carried over the wing, preferably close to the wing-span for reasons of center of gravity. So even though a plane may have a long cabin with a fifth seat at the end of the tunnel, that portion of the cabin is not useful for a large supply of fuel. Ferry tanks usually replace the back seat, or sit on the floor next to the pilot, while the right front seat is stowed in the back. A custom-made tank might fill the floor space behind the front and back seats, and cantilever out over the back seats; or the back seats might be removed and a large flat tank installed. One of the problems of equipping small planes with cabin tanks is getting the tanks into the airplane. The single-engine Cessnas are good in this respect, with their flat

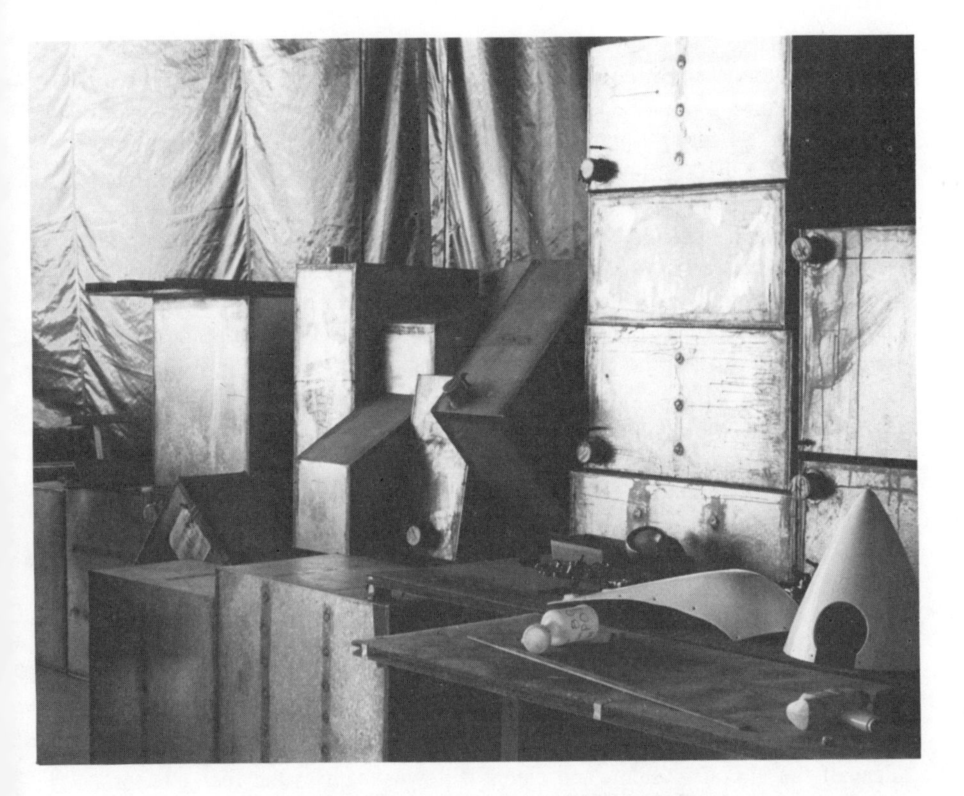

One-use ferry tanks made of galvanized steel for Southern Cross, a California ferry firm that specializes in group deliveries of Cessnas to Australia.

floors and large doors. Low-wing airplanes are less convenient. The normal solution is to make several small tanks, if one large one can't be gotten into the plane. In designing a custom-made tank, you should make a cardboard mockup and try fitting it into the plane before ordering the genuine article.

An A&P must be consulted, for several reasons. One is that the tank installation will require a Form 337 to make it legal both here and abroad, and a mechanic will have to approve the finished installation. Another is that few owners are aware of all the requirements for aircraft fuel tanks—venting, feed line size, and in some cases return lines. Continental fuel-injected engines return fuel to the tanks from the injection pump; the pump delivers more fuel to the engine than the engine needs, and the surplus is returned along with bubbles and vapor that may form in the fuel. Bendix systems, used on Lycoming engines, work differently; they bring to the engine only the amount of fuel it needs in the first place. Continentals may require special protocols for using the ferry fuel if return fuel goes to a single tank.

Several methods of hooking the supplementary tank up to the aircraft system are used. One is to plumb into the aircraft system between the fuel selector and the engine, and put an on/off valve on the auxiliary feed line. Another is to permit the aux tank to drain into one of the wing tanks and to feed to the engine using the standard airplane system. Still another is to pump fuel from the cabin tank into the airplane tanks, either with an electric or a manual pump. The simplest system is the first one, which is independent of the aircraft system, and enters it downstream of the fuel selector. The principal danger here is that you would forget to select "off" on the airplane system, and the fuel from the cabin tank would drain into the single tanks and vent overboard, or vice versa.

Fuel tanks must be vented so that a partial vacuum is not produced in the tank as fuel is removed from it. Fuel vents are usually forward-facing; on some high-wing Cessnas, for example, the fuel vent is a piece of tubing which curves down out of the underside of the wing behind the strut. It is placed behind the strut to prevent ice accumulation. That fuel tanks be vented to a non-icing location is of the first importance, particularly on the Atlantic route. An alternative to a forward-facing vent is one which faces directly aft in the wake of some portion of the airplane, such as the wing trailing edge or the back end of the fuselage. A vent should not, however, simply be cut off parallel to the skin of the airplane, because it might be exposed to an area of low air pressure, and tend to suck air out of the tank, rather than push it in.

So vital is an adequate tank vent that a cabin tank should be equipped with a second vent through a petcock inside the cabin. This vent would be kept closed to prevent fumes entering the cabin, but could be opened if for some reason the tank developed a problem with its overboard vent.

The fuel level in a fuselage tank can be judged in several ways. Ferry tanks usually have no fuel quantity indicators, and their capacity is thought of in terms of hours, not gallons. One simple arrangement is to build a piece of protruding tubing into the bottom and top of a visible side of the tank, and to connect them with a piece of transparent fuel line, which then serves as a standpipe.

A nice method of plumbing the cabin tank into the system is with transparent tubing. Not all transparent tubing is suitable for use with gasoline, however. One type that is is polyurethane flexible tubing and can be obtained from Ryan Herco, 2509 North Naomi Street, Burbank, California 91504. By using transparent tubing within the cockpit, you can see whether lines are full of fuel and whether bubbles are entering the system. Ryan Herco also sells flow indicators, which give a visible indication of the motion of the fluid passing through them. (Care must be taken to order and use materials that are compatible with gasoline.)

Since fuel feed problems constitute one of the greatest hazards of overwater flight, it is extremely important that supplementary systems be flight checked. All emergency conditions must be considered. For instance, suppose you have installed a fuselage tank in a fuel-injected high-wing Cessna. Normally, fuel feed is taken care of by the engine-driven fuel pump. If the engine-driven pump fails, there is an electric auxiliary pump which will keep the engine running. However, the factory-installed engine-driven pump is sized on the assumption that it will be feeding fuel from wing tanks, which are higher up than the engine and supply a standing head of fuel to start with. The cabin tank, especially the lower portion of it, will be below the engine, and will put more of a load on the electric pump. The Continental engine will only operate properly within certain limits of fuel pressure to the throttle body; these limits are given in the engine maintenance manual. It would be necessary to demonstrate, before one could have perfect confidence in the system, that the aux fuel pump is capable of feeding the proper pressure to the engine from the cabin tank. If it won't, then an additional supplementary pump must be added. Bendix automotive pumps are usually used; they are cheap, simple, and use twelve-volt power.

Most mechanics checking out a tank installation for a 337 approval would not take this reasoning into account, by the way; they would concern themselves with the adequacy of tank supports and connections. Official approval of a system therefore is not sufficient warrant of its safety; you have to reason out all the possibilities for yourself. Because of the difference in the requirements of carbureted and injected fuel systems, and between different types of injected systems, a ferry tank system removed from one airplane will not necessarily work in another without modification—though obviously the essential item is a fuel-tight container of the proper size; the plumbing can always be rearranged to suit.

Wing tip tanks are available for some airplanes, for instance Bonanzas, Comanches, and Navions. They hold about fifteen gallons each, and exact no performance penalty to speak of, since they slightly improve the span or span efficiency of the wing. Flint Aero, of El Cajon, California, sells supplementary tanks for high-wing Cessnas which, rather than being shaped like a salmon impaled on the wing tip, simply extend the span a little. They produce a larger span increase than does the other type of tip tank, and so should improve rate of climb and overall efficiency to some degree.

Tip tanks do not add enough fuel to make a big difference in an airplane's range, but they may make the difference between an airplane that can make Gander–Keflavik with ample reserves and one that must stop in Greenland. They do not increase the stress upon the wing spar, as cabin tanks do; in fact, they diminish it. They also reduce the amount of supplementary fuel that must be carried in the cabin. For a pilot who is planning to make a habit of long flights, or even one who is worried about fuel shortages here and there, the extra permanent capacity with no performance penalty is very desirable.

Oil capacity is less easily altered. Some ferry companies will not handle an airplane that has run out more than half its TBO, and part of the reason is oil consumption. An airplane owner usually has a fairly good idea of the oil requirements of his airplane, but prior to a long flight he should keep a log of oil levels and quarts added, and of the power settings at which the flying was done. Oil consumption will vary depending on the circumstances; high power settings often increase oil consumption (though some engines consume disproportionate amounts of oil at very low power settings), and turbocharged engines may use more oil at high altitudes than low. Oil consumption consists of two elements: oil getting past the rings and burning in the cylinders, and oil being blown overboard via the crankcase breather. The high oil consumption of turbocharged engines at altitude is usually the result of the latter, and can generally be corrected with an oil separator. Breather lines should be lagged (insulated) if flights in subfreezing conditions are planned, and a separator should be located on the hot side of the engine baffles and should also be insulated, in order to keep water vapor, which is always present in the crankcase, from condensing in the separator and returning to the case.

The minimum recommended oil level in pilots' handbooks is a conservative value, and ferry pilots simply fill (but don't overfill) the crankcase at the beginning of the flight and assume that the oil level will still be adequate at the end, even if three quarts or four are used in fifteen or twenty hours. An engine using more oil than this is suspect anyway; it may have broken rings or a crankcase leak.

It is possible to rig up systems to add oil in flight, and I suppose one could get them approved. The principal caveat is that oil must enter the crankcase below the level of oil in the crankcase; otherwise air flow brought about by the high pressure inside the crankcase will prevent oil flowing down the line at all. In practice, without modifying the sump, this would mean replacing the drain plug with a tee fitting with a drain on the run and an input line on the leg—a procedure which will work on some airplanes but not on others, because of interferences with nosewheel retraction or other obstacles in the engine compartment. I must say that I have never heard of anyone installing a supplementary oil system for a ferry flight, however long, and I don't think it is a very serious concern. If the engine uses so much oil that it can't make a fifteen-hour flight without running dangerously low, then it probably has no business making overwater flights in the first place.

If supplementary tankage and other special equipment push the weight of an airplane above the certified gross, a special permit must be obtained for over-gross operations. Permits are routinely granted for weights up to 10 percent over gross; up to 25 percent over gross might be approved with greater difficulty. Airplanes can of course carry more than that; Max Conrad used to take off at nearly 100 percent over gross, but he was a Piper factory pilot and could get approval for anything. Since you will in general want a 20 to 25 percent reserve over the still-air requirement for the leg you intend to fly, the distance you can attempt will be, say, 80 percent of your fuel weight multiplied by the number of miles you can expect to go per pound of fuel. Fuel weighs 6 pounds per gallon, and the airplane manual range figures will say, for instance, that with a 60-gallon fuel capacity the maximum range is 930 nm. This includes a 45-minute reserve; if fuel consumption is, for example, 10 gph (gallons per hour), then 930 nm can be flown with 52.5 gallons, or 315 pounds of fuel. The airplane will go 2.95 miles per pound of fuel. For a 2,000-mile leg, then, you will need about 680 pounds of fuel, and with a 25 percent reserve, 850 pounds. This is 490 pounds of extra capacity. Here you may run into problems if your airplane is unable to carry full fuel and full cabin load at the same time; but if it can, then, subtracting 340 pounds of back seat passengers from 490, you get only 150 pounds of excess weight (assuming that you will use the entire baggage capacity for baggage, survival equipment, and so on). This is obviously less than a 10 percent overload.

Since ocean flights invariably begin from coastal airports at low altitude and usually in mild or cool weather conditions, performance is not seriously compromised by a 10 percent overload, though the maximum rate of climb will be reduced by 20 or 30 percent, not by 10. A 25 percent overload, on the other hand, seriously diminishes climb performance. Twins require special consideration. Even a 10 percent overload will eliminate whatever single-engine capability light twins have. In fact, it is difficult to find a ferry pilot who does not express a preference for single-engine planes for this reason, if for no other.

Ocean flying is the profession of a few specialists, but it is by no means inaccessible to amateurs. In general, it is not even very difficult; as with most things that you think about and ponder a great deal before doing them, the execution turns out to be anticlimactic. All an ocean crossing really involves is getting into your plane, heading in the right direction for a long time, and then landing. Nine times out of ten, that is; it is that tenth time that requires all the preparation. You never know but what your one time in ten might come on your first try. Most of the problems of ocean flights are imaginary; by making adequate prior preparations, you can keep them that way.

Travels with *Melmoth:*

Scenes from Some Long-Distance Flights

Melmoth *in England in 1975. The nonstop flight from Gander, Newfoundland, to Shannon, Ireland, took 10 hours and 45 minutes, and consumed 106 gallons of fuel; the average speed was about 180 mph in zero wind. Although this was, to our knowledge, the first time a homebuilt had made the nonstop crossing, our arrival passed unremarked except by one alert reader of* Flight International, *who sent the magazine a snapshot of the airplane parked at Shannon. Touring Ireland, Scotland, England, and France by air proved quite convenient, and gave us a new and beautiful perspective on places we had visited before by car and on foot.*

Melmoth *over England. The choice of a short wingspan (23 feet overall) and tip tanks for a long-range airplane with a gross takeoff weight of almost 3,000 pounds was ill-advised. For optimum efficiency a much longer span would have been desirable; but even for a combination of a high cruising speed and good efficiency, a span 4 to 7 feet longer would have been better. At the time I designed the plane, I did not understand these relationships clearly. In spite of my ignorance, however,* Melmoth *is capable of nonstop flights of well over 3,000 miles, showing that a large fuel capacity and a powerful engine can make up for the deficiencies of a short wing in both range and climb performance. All 154 gallons of fuel are carried in the wing outer panels and the tip tanks; allowable cabin load is an additional 500 pounds. Empty weight—an embarrassing 1,500 pounds.*

Before our takeoff from Japan, a troupe of stewardesses from the domestic airline, All Nippon Airways, asked to be photographed with Nancy. Customs and Immigration officials looked on, as did the plainclothes policeman we called "Number One" (second from right), who came out to give us our pistol back. The .22 caliber handgun, a piece of survival equipment for the Alaskan portion of the trip, created an 11-hour stir among officials at Chitose, and qualified us as genuine Americans. We made several newspapers and a TV talk show, and got to eat a lot of free sushi.

Aleutian weather from above: volcanic cones emerge from a boundless blanket of fog. The Aleutian chain provides some scenery and a few navigational facilities on the way to Japan, but few landing places; its scarce airstrips are all military and closed to civil traffic. One, Amchitka, looked promising at first but turned out to have been closed on account of radioactive contamination from an underground nuclear test. What looks like a piece of cellophane tape running up the windshield is the antenna of the AM/FM radio; it permitted us to listen to Chopin on a Siberian radio station from somewhere hundreds of miles off the Russian coast.

The full panel in 1976, photographed on the return flight from Japan to Alaska. The ADF is tuned to the Nikolski beacon, between Adak and Cold Bay; TAS is 186 mph at 21 in. Hg and 2,500 rpm. Except for running at peak EGT, which I always do anyway, I made no attempt on ocean flights to employ maximum-efficiency speeds and power settings, because the airplane's fuel supply was so ample. On the right side of the panel, below the circuit breakers, is the HF radio; above it and to the right, hidden in shadow, the invaluable fuel flow totalizer. The autopilot, as always, is on. After flying 11 hours overnight from Chitose to Adak, being detained 5 hours at Adak, and then setting off on the 7-hour flight to Anchorage without rest, we both fell asleep in the air. Nancy awoke, noticed that I was asleep, but was too tired to care, and went back to sleep without awakening me. Fatigue can affect you in strange ways.

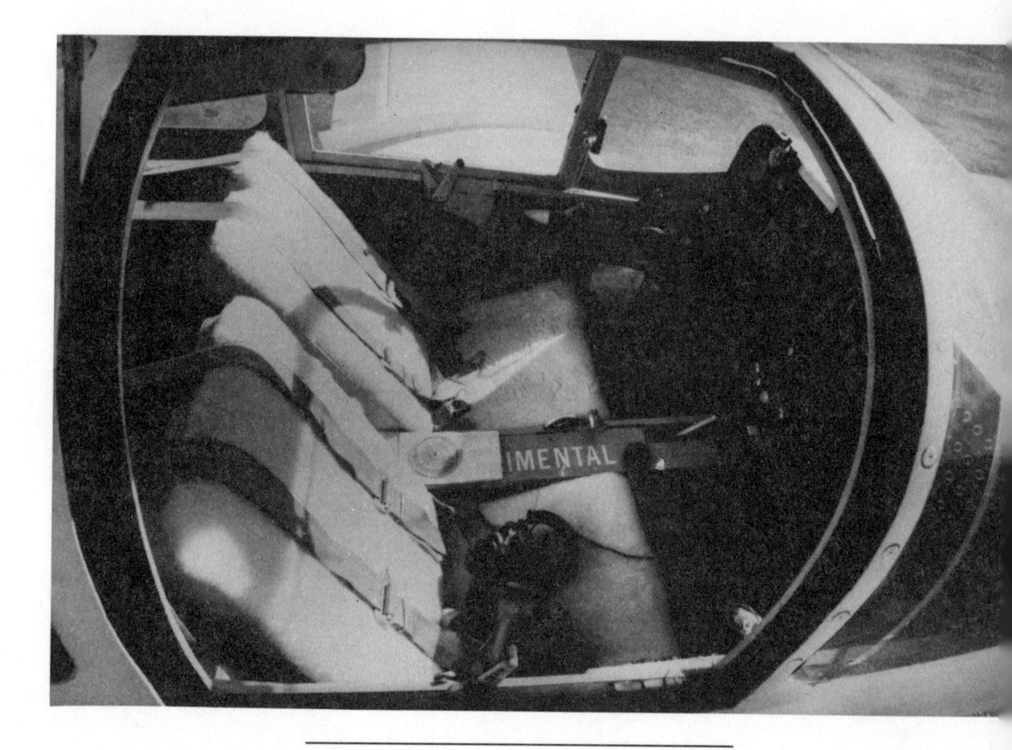

Comfort is an important issue in a long-distance airplane. After making trips to Europe and to Japan, I replaced the dismal fixed seats in Melmoth *with new ones having adjustable backs and sheepskin covers. These I later modified (after this picture was taken) to incorporate lumbar support, which was very helpful, and a longer thigh support, as well as a variable tilt of the seat pan. The ability to change positions is valuable; the console between the seats and the nosewheel-well further forward are a problem in this respect, and airplanes with flat cabin floors which enable you to stretch your legs diagonally across the cockpit are preferable. ''IMENTAL'' is part of a required passenger warning on all homebuilt craft: ''EXPER'' are the hidden letters.*

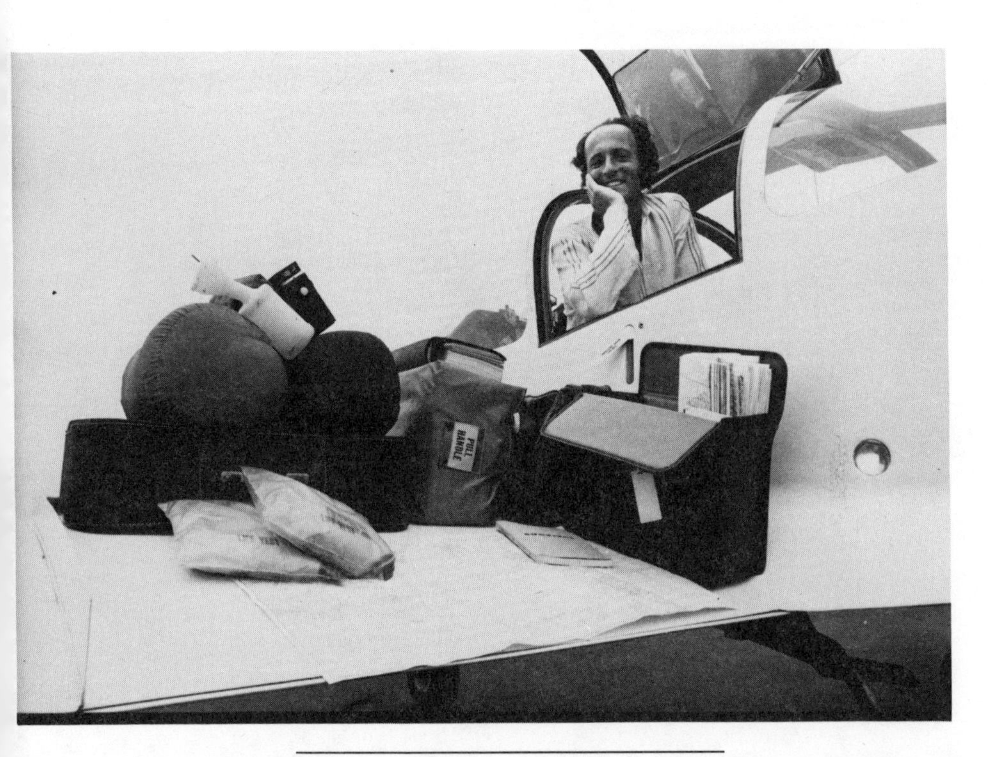

The author and his transatlantic baggage. Besides two suitcases, the ocean flight required a life raft, two life jackets, a bundle of charts, a portable direction-finding radio, a survival kit, an Air Force survival manual, a portable urinal, two sleeping bags, and a couple of thick novels (not shown).

4

NAVIGATION

Within a land area, like the United States, the navigational problem takes the form not of trying to ascertain at all times where you are, but only of selecting a route. The vast supply of radio navigational aids makes knowing where you are perfectly easy. But a long-distance flight differs from a short one in several respects. There is, to begin with, the fact that over a long leg the effects of the spherical shape of the earth begin to be significant, and the shortest distance between two points—always a great circle—begins to differ significantly from the rhumb line, or straight line distance as it would be plotted on a Mercator projection.

For flight planning we usually use charts drawn on a ''Lambert conformal conic projection,'' on which deviations from great circles are smaller than they would be on a Mercator chart. The Mercator is quite useless, in fact, except as a wall decoration in schools, because of the absurd distortions it introduces far from the equator, making Antarctica appear to be the largest land mass on earth, and Greenland rival in size Australia and South America.

The ideal for a map is to accurately represent the distances between any two points on it. The degree of deviation from that ideal is proportional to the degree of difference between the flat plane of the map and the curved surface of the earth; a map purporting to show half the earth's surface will contain sizable errors, while one showing fifty or a hundred square miles will differ little from the curved surface.

On the conic projection, lines of latitude—the ones which run parallel to the equator—are kept parallel to one another, but are curved; the lines of longitude, or meridians, are held straight, but allowed to converge. The effect is that of taking the portion of the earth's surface represented by the map and flattening it onto the surface of a cone—a cone which possesses a curvature most like the piece of geography in question—and then unrolling the surface of the cone and laying it flat upon the table. This projection is well suited for charts depicting relatively small areas; as with other flat maps, it becomes less accurate as the area depicted gets larger.

One solution to the problem of plotting courses over long distances is to plot them upon a globe, select a series of landmarks or waypoints which fall at intervals along the great circle, and then transfer these to some more convenient chart, where they will determine segments of a curved line. A convenient globe for this purpose can be obtained from the National Geographic Society, along with a hemispherical plastic cap which is calibrated in miles, kilometers, and degrees, and which can be used for plotting and measuring great circles. However, globes are intended for illustration and education, not for navigation, and they are not overly precise.

The size of the difference between the great circle and the straight line or rhumb line track varies with its position upon the earth and its length. Tracks close to the equator, and those running nearly north and south, differ little; those running at large angles to the meridians and those located far from the equator may differ greatly. The greater the distance to be covered, the greater the difference, if there is one, will be.

Since people customarily visualize the world in a Mercator projection, and think in terms of north, south, east, and west as though the world were laid out in a grid like Manhattan, they may find it difficult to accept the realities of great circle flying. Most people, furthermore, never have occasion to disabuse themselves. They automatically suppose, for instance, that since Japan is far to the west of California, and Hawaii is in the middle of the Pacific between the two, Hawaii is a natural way-station on the route to Japan. They are surprised to discover that the direction to fly to get from Los Angeles to Tokyo is not westward but northwestward, up the coast, and that the shortest distance between the two cities—the real "straight line"—grazes the Aleutian Islands. Anchorage, Alaska, is a more natural stop between Los Angeles and Tokyo than is Honolulu.

There may be reasons for flying to Hawaii on the way to Japan, but

The National Geographic Globe with its mileage-measuring plastic cap.

directness is not one of them. The great circle distance from Los Angeles to Tokyo is about 4,730 nm; via Honolulu and Wake Island, it is almost 5,920 nm—a difference of almost 1,200 nm, or 25 percent of the great circle distance, time, and fuel burn. On the other hand, a hypothetical waypoint at Anchorage increases the great circle distance by only 280 nm or so, and one at Cold Bay or Adak by somewhat less.

Even for shorter routes, the great circle may be attractive. Staying for the moment with the trip from Los Angeles to Tokyo, look at the final long overwater leg from Alaska to Japan. You would probably take off from Cold Bay, which is the westernmost civil airport with a good long runway to be found in North America. (Adak and Shemya are difficult for the reasons I have mentioned before—weather and the reluctance of the military authorities to permit civil aircraft to use them.) From Cold Bay you would have the choice of striking out along the great circle to, say, Chitose Airport on Hokkaido, the northern Japanese island, or of following the chain of the Aleutians out to Shemya and then taking a cut to the southwest. The great circle distance from Cold Bay to Chitose is 2,246 nm; the total length of legs using waypoints at Adak, Shemya, Noho (a waypoint off the Japanese coast whose purpose is to keep you out of Russian airspace), and Chitose is 2,456. The difference of 210 nm might represent almost one and a half hours' flying time, and a corresponding price in fuel. In compensation, following the island chain to Shemya offers much simpler navigation for much of the trip, since there is a VOR at Shemya, an NDB at Adak, and there are various facilities on Hokkaido.

The great circle, which passes north of Adak and south of Shemya, could be used if a little more effort were put into navigation. It would be useful to have DME aboard for a good position and speed fix off Shemya. On this particular route the possibility of a perfect great circle is at any rate spoiled by the Russians, who overhang Japan like an eave and force incoming traffic from the north to swing eastward to avoid them. For a flight from Anchorage to Tokyo they are not a concern, but a flight from Cold Bay to Kushiro or Chitose has to make the dogleg at Noho, or slightly west of Noho.

The same kinds of considerations apply to any trip. The North Atlantic, which is the body of water that sees the largest share of light aircraft flying, offers a number of different routes, depending on the capabilities of the airplane. Like the Pacific great circle from California to Japan, the Atlantic great circle grazes land throughout much of its length from New York (say) to London or Paris. The stretch between the North American continent and Europe can be taken in a single bound (Gander to Shannon) or in a series of hops, from Labrador to Greenland to Iceland and so on. A route requiring somewhat longer range than the northern stepping-stone route, but offering better weather, is that via the Azores. From Gander to Shannon is 1,715 nm; via the northern lands the distance becomes 2,732 nm, but the longest leg—that from Goose Bay to Sondre Strom—is only 866 nm. Flying via Fort Chimo, a still more remote outpost in Labrador (often used by jets en route from Wichita to Europe, or headed for Scotland rather than Ireland) all the leg lengths can be brought down under 725 nm—the range, without additional fuel, of many light airplanes. Each deviation in the cause of shorter leg lengths, however, increases the length, in miles and days, of the entire journey.

Long legs are preferable to short ones for a number of reasons; neurotic haste is only one of them. The transatlantic trip of Giorgio Toscani in a Cessna 152 is instructive. Without ferry tanks, HF radio, or any of the other customary appurtenances of the long-range traveler, Toscani set off from Canada in September of 1978. The advertised range of the 152 is 670 nm at 55 percent of power, at a cruising speed of 90 knots or so. He was consequently obliged to stop at every microscopic outpost along the way to refuel and rest. He headed out of Moncton, New Brunswick (the place where pilots stop to put their papers and permissions in order with the Canadian government before starting off on a transatlantic trip), toward Fort Chimo; because of strong headwinds, he had to stop halfway to refuel. But he reached Fort Chimo by evening. Then weather obliged him to remain for several days at Fort Chimo, where the only hotel cost $96 a night, before continuing to Frobisher on Baffin Island. There he spent two more days, delayed first by a dead battery and then by more weather. Finally he crossed his first long stretch of open ocean, from Frobisher to Sondre Strom. There he again waited several days for weather before continuing to Kulusuk, a desolate island on the east coast of Greenland, where bad weather closed in on him again and his airplane was damaged by violent winds. An auto mechanic repaired the plane; but then the runway remained closed for several days because of snow. Finally he got to the far side of Iceland, whence, after a couple of attempts aborted because of weather, he reached Vagar, an island in the Faeroes north of Scotland, and from there, finally, Scotland itself.

"The trip was an incomparable experience," he wrote, "though I don't know if I'd recommend it."

Toscani's main problem was that he was underequipped—obviously—and that the circuitous route that he was forced to take led him into areas of notoriously bad weather. In order to appreciate the significance of his long waits for weather at a series of unpleasant places, you have to consider that he never knew that he would in the end successfully complete the trip. Winter was coming, his progress was slow, and he was beset by every kind of difficulty. Furthermore, the quality of life at those dismal outposts in northern seas is not such as would make you happy to have lingered upon them. He must have spent weeks in the blackest of moods, relieved only by an occasional success, and ending of course in the elation of finally landing in Europe.

Don Taylor, the retired Air Force meteorologist who flew his homebuilt Thorp T-18 around the world, encountered different yet comparable problems in crossing the Pacific. Taylor's 2,000-mile range was much superior to that of Toscani's 152, but it was still inadequate for Pacific distances. Taylor's plan was to island-hop to Midway and then track due north into Adak—an immense dogleg in the interest of avoiding the 2,200-nm leg from Honolulu to California. He was delayed for nearly two weeks by U.S. Navy authorities at Midway, where he landed without the proper authorizations. It took the intercession of some congressmen to make the Navy, whose mood can be like that of a Greenland cold front, release him—though once it did let him go, it went so far as to send a patrol airplane ahead of him to Adak to check the en route weather.

Taylor had attempted the same trip two years earlier, that time planning to fly from northern Japan to Shemya at the tip of the Aleutians; but he had been delayed at various points on the way around the world to Japan, and when he was ready to cross the Pacific, the weather had turned bad for the winter. He had his airplane shipped home in a crate. Another pilot, an Australian with little experience, flying an American Yankee equipped with a spare tank, essayed the same route, but landed in Russia with a fuel-feed problem, real or feigned (one never knows). After a stay of a couple of days during which he was hospitably entertained and also refueled, he continued on his way.

Yet another pilot left Kushiro on Hokkaido on a flight plan for Cold Bay, a destination which he could not reach, not having sufficient fuel. He reportedly planned to contrive an emergency and land at Shemya. Unfortunately he missed Shemya completely, and was finally obliged to ditch his Piper Super Cub somewhere off Adak, where he was rescued by the same Navy which would not have granted him permission to land. His airplane was wrecked in the process of being winched out of the water—this always seems to happen, because of the weight of the water in the airframe—and he was very ill-tempered with his saviors as a result.

High adventure is usually highest in retrospect; and the adventures of long-range flying generally consist, while they are taking place, of a kind of tense boredom. It is understandable that the pioneering transoceanic pilots made their attempts with inadequate, or barely adequate aircraft; there was nothing else. Today, it is not so difficult to increase the range of an airplane to the point where it can make an ocean crossing with comfortable reserves, and it is the better part of valor to do so. When every stop is full of pitfalls, it is best to stop as rarely as possible.

One of the pitfalls of stops in remote outposts, or even less remote ones such as the Azores, is the costs they can entail. Landing fees and miscellaneous fees of over $100 are common; overnight lodging in hotels may be similarly expensive; fuel may sell for $5.00 a gallon or more. At such prices, a more expensive ferry tank installation giving more range may turn out to be an economy. Sometimes fuel taxes may be deducted or reimbursed when the fuel is used to leave the country in which it is bought, but sometimes the price of fuel is simply astronomical. Fees for using radio facilities may also be high; after crossing the Atlantic, for instance, you seem to keep getting bills for months afterwards. Stopping as rarely as possible also means flying the shortest total distance, which in turn means being able to use great circle navigation techniques.

For purposes of navigation, the earth is represented as a perfect sphere of a certain diameter. Locations upon its surface are described by latitudes (the horizontal lines encircling the globe parallel to the equator, also called, for this reason parallels) and longitudes. Latitudes run from 0—the equator—to 90 degrees north or south; longitudes (or meridians) are numbered east and west from an arbitrary one through Greenwich, England (a suburb of London and the location of the Royal Observatory), to 180 degrees. At 180 degrees west and east meet, somewhere in the middle of the Pacific Ocean.

It is not always easy at first to remember which is which; you might try saying to yourself that all longitude lines are equally long, whereas all latitude lines are not. When the coordinates of a location are given, latitude is invariably given first; and the numerical value of the latitude is always followed or preceded by N or S, and that of the longitude by W or E. For purposes of aerial navigation, latitudes and longitudes are expressed in degrees, minutes (each minute is 1/60 of a degree), and tenths (or hundredths) of minutes; *not* in degrees, minutes, and seconds. To give a sense of the scale of these values, it is useful to know that one nautical mile is equal to one minute of arc upon the earth's surface. That is, one minute of latitude; because latitude is measured north–south *along* the lines of longitude, which are all the same length. One minute of longitude is one nautical mile at the equator; but it gets smaller and smaller as the distance from the equator increases. About 200 yards from the North Pole, one minute of longitude is only one inch long.

Since the earth rotates once—360 degrees—in 24 hours, it turns 15 degrees per hour. Thus each of the time zones around the world is 15 degrees wide. Political boundaries and practical reasons dictate that time zones are not outlined by straight lines of longitude, but instead by zigzagging lines which skirt continents and states. But like intervals of longitude, the time zones grow narrower close to the poles; in Alaska, between 60 and 70 degrees north latitude, a time zone is only 300 miles wide, and a fast airplane can make the sun stand still. This seemingly insignificant piece of information can in fact be quite important in flight planning, because the rate at which one crosses time zones will influence the amount of daylight one can expect during a flight or, conversely, the amount of darkness. On the occasion that I flew to Japan from Alaska, I had spent the previous week in Alaska and had grown accustomed to summer nights four hours long. When I took off at nightfall for Japan, quite tired already, I expected help from the sun, which I thought would rise in four or five hours, in staying awake. I forgot that I would be flying across a number of time zones; in fact, the darkness lasted nearly thirteen hours.

It is also near the poles that the difference between a great circle and a rhumb line course is greatest. A rhumb line course is defined as one that crosses all meridians, or longitude lines, at the same angle; on a Mercator projection it would be a straight line. As a practical matter, there are few routes that you would be likely to fly above 65 degrees north latitude, and none whatever below 45 degrees south, unless you were, for some unimaginable reason, undertaking a transpolar flight. (There is only one transpolar route that goes from a plausible origin to a plausible destination: that is the one over the North Pole from Point Barrow, Alaska, to the northernmost parts of Norway. Any other origin in North America lands you somewhere in the Soviet Union, assuming that you want to cross the pole en route and not on a dogleg performed purely for the sake of crossing the pole. The South Pole, because of its remoteness from all other continents and the fierceness of its weather, is not even worth considering.)

At any rate, the difference between a great circle and a rhumb line is greatest near the poles, where, as you can figure out for yourself with a little reflection, the worst case is one in which the rhumb line is $\pi/2$, or about 1.57 times the great circle. At 65 degrees north, two points 2,000 nm apart on the great circle are 2,160 nm apart on a rhumb line. The ratio increases with increasing distance, but 2,000 miles is typical of the leg length that a light airplane might make.

The difference between the great circle and the rhumb line is greatest far from the equator; when the difference in longitude of the origin and destination is great; and when the course is aligned more with parallels than with meridians.

As I mentioned early in this chapter, many miles can be saved by following great circle routes. It may not seem so from the comparison between rhumb line and great circle routes that I just made. But the difference is in the total distance to be flown. A given leg may be longer by only a small percent along a rhumb line. But if you are flying, say, from California to Greece, the great circle is more than a thousand miles shorter than the rhumb line.

The value of the rhumb line to pilots is that it permits them to fly a constant heading, whereas the great circle requires, in theory, constant changes of heading. In practice, even when flying a great circle, the pilot reduces it to a portion of a polygon, making heading changes of one or two degrees at intervals of an hour or so, and thus converting the great circle into a series of rhumb lines, just as, in initially planning his route, he would determine, from a globe, the great circle track, and then select refueling and rest stops distributed along it. As a practical matter, magnetic variation may make it necessary to change heading several times even along a rhumb line; or, conversely, as on the Pacific route between Shemya and Hokkaido, it may coincide so precisely, but inversely, with the changing great circle heading as to make heading changes unnecessary.

A route like that from Gander to Shannon is customarily broken down into rhumb line segments, each bridging 5 or 10 degrees of longitude. A magnetic heading is calculated for each segment, and corrected for the forecast or observed winds. That route, which is mostly bracketed between 50 and 55 degrees north, crosses 5 degrees of longitude in about 200 miles; and so every hour or ninety minutes or so—depending on the speed of the airplane—the pilot can look forward to a refreshing little set of routine chores. He checks his directional gyro against his compass, changes his heading to the new precomputed value, and calls Gander or Shanwick on HF to report his position which, of course, is merely an estimate unless he is equipped with Inertial, VLF, or Loran. He then settles down to another hour of inactivity.

Most ocean flights are made without the benefit of long distance navigational equipment, because most such flights are delivery flights, and the airplanes are not ordinarily equipped with the necessary radios. There are three different systems which are suitable for long-range nav, omitting celestial navigation, which is impractical in light airplanes because of the limited visibility from the cockpit and the distortions introduced by the windows. The most elaborate and complex is Inertial, or INS, which is used by most international airline aircraft. Inertial is a dead reckoning device. On the principle that if you have been flying at a certain groundspeed in a certain direction for a certain length of time you can be in only one place, it continuously computes the direction and speed of the aircraft and determines where it is, displaying that information digitally, in latitude and longitude. It determines speed and direction inertially; that is, precise gyroscopes and accelerometers measure every change in speed and direction of the airplane without reference to any outside system, such as radio stations. By summing up all of these movements, the system is able to compute the position of the airplane on the earth with accuracies of a mile or two over transcontinental distances.

Obviously, extreme precision of manufacture is necessary for such systems, and their cost, to say nothing of their size and weight, is prohibitive for personal aircraft, except the most extravagantly equipped jets and turboprops. New, smaller, lighter, and cheaper Inertial systems are being developed, some using laser beams in place of gyroscopes and accelerometers; but the day of the "strap-down" system for light aircraft—if it ever comes—is still far away.

Another type of system, VLF/Omega, behaves like an Inertial system as far as pilot actions are concerned, but functions altogether differently. As with Inertial, VLF systems must be informed before the flight begins of their position on the earth and also of the correct time. Internal computers rectify the time and position information, and the system locates itself on a grid which covers the entire earth and is composed of very low frequency radio waves (hence the term VLF). Eight sending stations, each with a range of up to 10,000 miles, provide the signals, which propagate like ripples in a pond. Whenever three or more stations are available, the system is able to keep track of its position by noting how many "lanes"—that is, wavelengths—of each station it crosses, and in what direction. The method has been compared by writer Peter Lert to a fly making its way along a globe over which a nylon stocking has been stretched, and knowing its location by counting the number of threads it has crossed, in each direction, since it started. By knowing the time of day and the date, the system's computer can make allowances for seasonal, diurnal, and geographical distortions in the signal mesh.

If it loses the signal from some of the stations it is using, or if the signal quality drops below a certain minimum level, the VLF system shifts into a dead reckoning mode, in which it continues to update the airplane's position based on speed and direction of flight. Heading and true airspeed (TAS) information may come from HSI and air data computer inputs to the VLF/Omega computer, or TAS may be manually entered. Wind will be based on the last measured value.

There are a handful of VLF/Omega systems on the market, manufactured by Litton, Bendix, Tracor, Communications Components Corp., and Global Navigation. Their prices start at around $30,000, varying somewhat with the type of installation and the customer's requirements, as well as the inclusion of certain optional equipment.

Both Inertial and VLF/Omega systems are forms of what is called "area navigation." Area navigation, as I mentioned in an earlier chapter, means navigation that is independent of specified airways or routes, and that is able to use waypoints selected at will by the pilot. In other words, it is navigation along any route, rather than certain preordained routes. Area navigation contrasts with airways navigation, which takes all traffic down certain airways whose locations are invariable and which are determined by the positions of

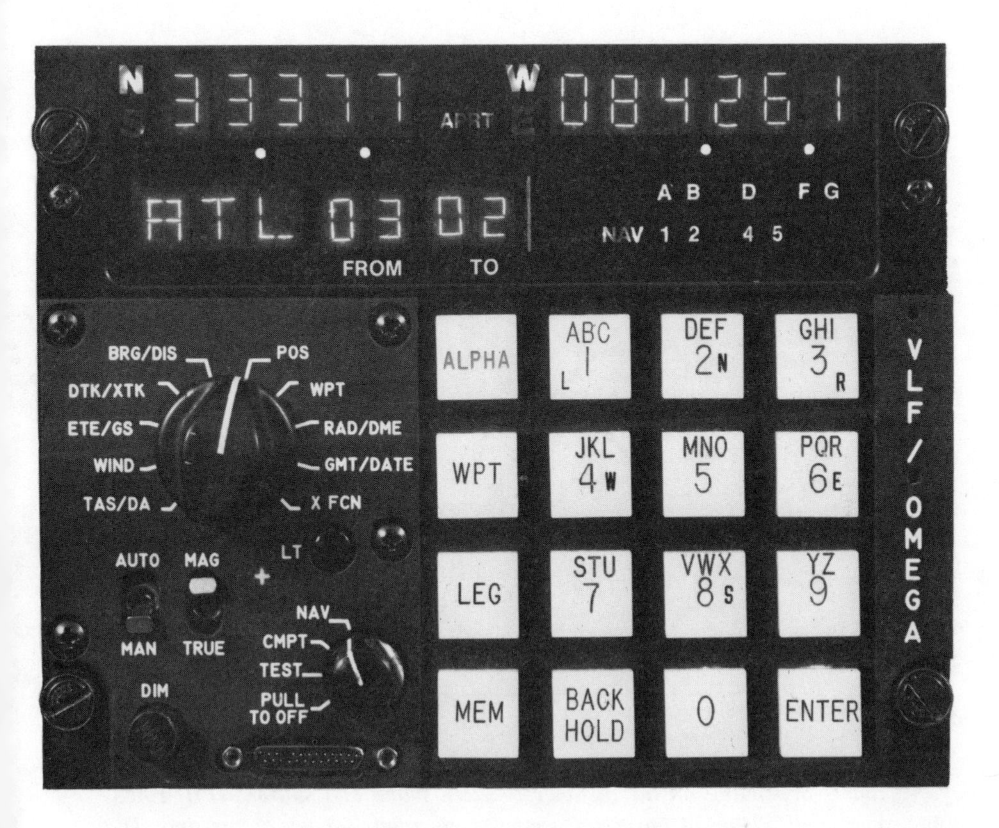

The Control Display Unit—CDU—of the Collins LRN-85, one of the most sophisticated and expensive VLF/Omega systems. The keyboard and the selector switch used to select the information displayed are typical of this class of device. Courtesy Rockwell International.

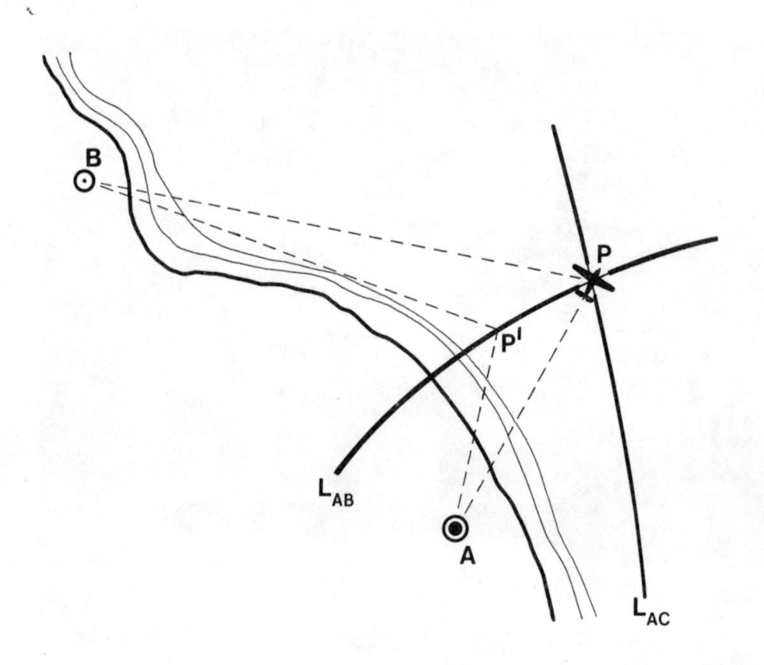

How Loran works. A "master" station (A) and two "slaves" (B and C) broadcast simultaneous pulses, which arrive at the airplane (P) at different times, depending on the distance between plane and stations (PB, PA). A hyperbolic "line of position" connects all points for which the time delay between the arrival of pulses is the same; that is, all points for which PB-PA = P'B-P'A. Two lines of position, one from each slave, determine a point. Some modern equipment, like the TI9900, convert the result into a lat/long display.

VOR stations on the ground. Inertial and VLF are unique in offering worldwide navigational coverage. The Omega system was in fact first developed by the U.S. Navy to assist submarines in navigation. It is said that a VLF radio can work anywhere, even at the bottom of a well. Both VLF/Omega and Inertial are of limited interest to the light-plane pilot, however, because they are so costly.

Another long range system is Loran, which was developed during World War II. The name stands for "long range navigation," and until the relatively recent introduction of VLF/Omega and Inertial, Loran was the most accurate system available (apart from celestial) for navigation at sea. Loran chains consist of pairs of stations, called masters and slaves, which broadcast synchronized pairs of pulses. Depending upon the location of the aircraft (or surface vessel) between the pair of stations, the signals will arrive at slightly different intervals, since one will take longer to reach the receiver than will the other. Using delay circuitry which synchronizes the two signals and notes the time shift necessary to do so, the Loran set identifies a hyperbolic "line of position" on which the receiver is located. Since time discriminations can be made to accuracies of better than a ten-millionth of a second, during which the electromagnetic signal travels less than a fiftieth of a mile, the accuracy of a line of position can be quite good. A second line of position from a second slave station is necessary to provide an intersection of two lines, and therefore a point location for the receiver. However, in many cases a single line of position, combined with dead reckoning, can give useful position information.

An analogous system called Consolan exists, but it is little used. One ferry pilot once commented to me that Consolan was a good way of finding your position, if you already knew where you were. Consolan broadcasts a low frequency signal (around 200 kHz) which is audio modulated through a circular sweep, somewhat the way VOR is modulated in the RF (Radio Frequency) band. Depending upon the listener's bearing from the station, he will hear first a silence, then a rising series of dashes, then a rising interference culminating in the disappearance of the dashes and their replacement by a continuous tone, and then the falling away of the continuous tone and its replacement by dashes, and finally a silence. By counting the dashes before and after the continuous tone and consulting a chart, he can determine his approximate bearing from the station. Consolan can be hard to use because of the lack of clarity of the received signal and the blurred boundary between the dashes and the continuous tones. It is like the four-course radio ranges of prewar navigation; better than nothing, but far from satisfactory. The San Francisco Consolan station, which might be useful on a flight to Hawaii, transmits on 192 kHz, which is below the receiving range of most ADFs. I have used two European Consolan stations to get a line of position between Ireland and Iceland, but as my cynical informant said, if I had not already known more or less where I was, the Consolan information would not have seemed very persuasive.

ADF can be surprisingly useful over long distances; sometimes you can receive the BBC, which transmits near London on 400 kHz, after your climb to altitude out of Gander. But reception is not dependable. It is better at night than during the day because of the reflective properties of the ionosphere, which forms a kind of waveguide with the surface of the earth and steers signals around the curvature of the earth. Everyone has had the experience of picking up on a car radio at night a station hundreds or even thousands of miles away. This effect, too, is unreliable; and except at close ranges the best you can expect from an ADF is a general reassurance that your destination is ahead and not behind. Coastlines can disturb even this level of accuracy; flying along the eastern coast of the Soviet Union, I had the depressing experience of receiving a broadcast station quite clearly, but of seeing its bearing off my right wing undergo no perceptible change over a period of half an hour, until the signal faded from sight. If there had been a clear swing of the needle, it would have been possible to know not only my bearing from the station, but even my distance from the station, given the rate of swing and an estimated groundspeed. Except close in, however, NDBs are rarely so cooperative.

Several ferry services have begun using marine Loran equipment for finding position, even though it is not officially approved for aircraft. *They* approve it. Product developments in this field are fast-moving, but as I write this the set of choice is the Texas Instruments TI9900, a Loran navigator about the size of a large phone book and weighing ten pounds. Within the areas of the world covered by Loran chains (which include the western United States, western Canada, Alaska and the Aleutians, Japan down to the Philippines and Southeast Asia, the Hawaiian islands and most of the Pacific between them and California, the eastern half of the United States, and the entire North Atlantic and Mediterranean), this little box gives position in lat/long to a tenth of a minute of arc (about 600 feet), and even incorporates a digital course deviation indicator which shows displacement from a rhumb line course in miles and direction to steer to return to the course. It can be programmed with waypoints, and it can compute a series of great circle waypoints between two points. It is also a dead-reckoning computer, displaying speed and ground track in relation to true course, time to go to the next waypoint, and so on.

The TI9900 was introduced at a price of $3,500; I believe this price will come down, however, because the device, for all its capabilities, is only a marriage between that company's earlier TI9000A Loran, which now sells for $1,500, and its TI58 calculator, which sells for under $100.

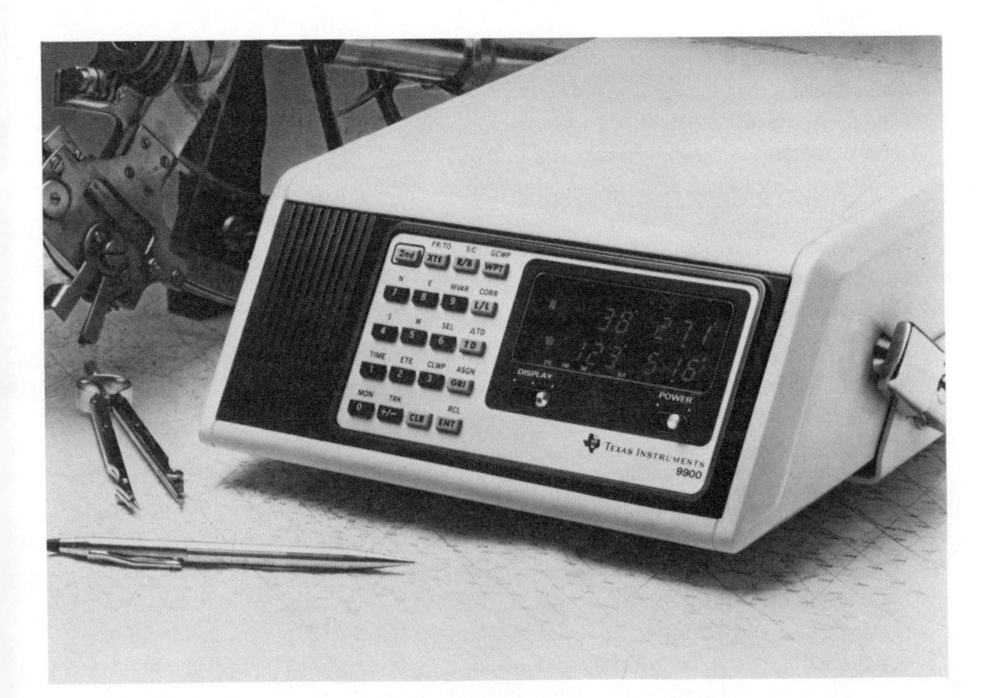

TI's latest marine Loran, the TI9900. About the size of a big phone book, it is not officially approved by anybody for use in aircraft, but it will work, and it represents a huge improvement over dead reckoning. Its $3,500 initial price tag will probably decrease with time. Courtesy Texas Instruments Inc.

Although it lacks the convenience of the TI9900, the TI9000A/TI58 combination can do most of the same things, and its aggregate price makes it more accessible for occasional use in overwater flying.

The ability to calculate the locations of waypoints along a great circle and to display them as lat/long positions, or to compute the distance between two lat/long positions, or to compute course deviations, seems quite magical; the only thing that comes close to it in everyday general aviation equipment is VHF Rnav, which still lacks the regal objectivity of thinking about position in terms of latitude and longitude. But these are all functions which can be performed with a hand-held programmable calculator, and there are several of these on the market which can be useful for pilots, especially for planning long flights.

The two principal competitors in this area are the Texas Instruments TI58 and TI59, and the Hewlett-Packard HP-67. Either can be used to perform rapidly the complex mathematics of route design on a spherical surface. The HP, which is more expensive, talks in RPN (Reverse Polish Notation) and seems to appeal to the professional scientist; the TI talks in colloquial algebraic idiom and aims at the wider consumer market. I have a TI59 with a cradle printer, a $350 combination which is very useful for flight planning, and so, with the observation that the HP can do the same things, I will describe some of its uses.

Although a globe is the best way to visualize the relationship of places on the earth to one another and to plan the broad outlines of a trip, a programmable calculator like the TI59 is a convenient way to plan a trip in detail. Prestored aviation programs are available; they supply precise answers to questions about heading and distance which can be difficult to answer solely on the basis of a chart. Charts are still necessary, both because even ocean flights begin and end with recognition of land masses or at least radio facilities, and because they are the source of magnetic variation information. But a calculator makes short work of the problem of working out a flight plan for a long route.

Three programs in the TI59's aviation module are specifically germane to long-distance flying. They are AV-04, Long Range Flight Plan; AV-10, Rhumbline Navigation; and AV-11, Great Circle Flying. Certain others, such as AV-07, Wind Components and Average Vector, and AV-02, Flight Plan with Wind, are also useful. The TI59 can be used with an optional cradle printer, which records results on a paper tape. The Long Distance Flight Plan program requires the use of the print cradle; for other programs, it is a convenience, permitting you to record data as the calculator produces it. For the programs not using the printer, the cheapter TI58 can be used, as well as with the TI9000A Loran set.

The TI59 is not difficult to use. Having inserted the aviation module, you

The TI59 and its companion printer: navigation in a nutshell. Courtesy Texas Instruments Inc.

call up the program you want to use by an identifying series of keystrokes. The calculator has available ten different labels (designated A, A′, B, B′, etc.) which are used by each program to identify certain data or operations. For instance, A and A′ may be the latitude and longitude respectively of the origin, B and B′ those of the destination; C could be the magnetic course, D the distance, and so on. These labels work two ways; you either enter a number and then press the label button, in which case you are telling the calculator, for example, that the latitude of the destination is such and such, or you press the label and the calculator displays the corresponding value, in which case it is telling you that, for instance, the distance between the two points you have entered is such and such.

The Long Range Flight Plan program takes the coordinates of an origin and a destination, plus those for a number of waypoints in between (as many as forty-four waypoints can be used), as well as groundspeed, fuel aboard, hourly fuel consumption, and estimated time of departure. Then, in the cute way computers have, it proceeds to print out reams of information about the trip.

For example, let's say we have before us Jeppesen's NP(H/L)1 chart: North Pacific High/Low Altitude Enroute. We want to assess the merits of several possible routes to, say, Japan.

We know from the globe that the most direct route is via Alaska and down the Aleutian chain; but we know from rumor, or from our climatological studies, that the weather there is likely to be very bad for most of the year, and quite bad for the rest of it, and that in spite of the longer distances, a route across the central Pacific might be more dependable. Don Taylor, having in 1973 abandoned his circumglobal record attempt in Japan because of bad weather in the Aleutians, chose on his second effort, in 1976, to island-hop across the Pacific. His range was about 2,000 miles, and he flew from Davao in the Philippines to Yap, Guam, Truk, Ponape, over Eniwetok to Wake, Midway, and thence to Adak. He was motivated in his island-hopping by a desire to keep his overwater legs as short as possible. If he had been coming from Japan, he might have gone from Tokyo to Iwo Jima, thence to Minami-Torishima, and thence to Wake and Midway. At any rate, he could have compared all these routes quite conveniently with the Long Range Flight Plan program.

This program consists of two parts: the first computes the overall distance, time en route, time of arrival, fuel required, and fuel left at the end, for the great circle between the first and last points on the trip. This information is of no interest when several waypoints are involved. The second part computes the same information for each leg.

First, you will call up program AV-04 by a series of keystrokes, namely, (2nd) (Pgm) 04 (2nd) (Deg) (SBR) (CLR); (the code may look puzzling, but it is simply a matter of pushing those eight buttons in succession). Then you enter the latitude and longitude, in that order, of each waypoint, beginning with the point of departure and ending with the final destination, labeling each of them "A" (the program takes care of numbering the successive waypoints). You then enter your estimated groundspeed for the entire trip (you can calculate individual legs if the groundspeed would differ greatly from the average; but since this whole procedure is only for rough trip planning purposes, you needn't); fuel aboard at takeoff (this can be any fictitious number, if the total fuel requirement exceeds the capacity of the airplane and refueling stops are planned); the fuel burned per hour, in gallons or pounds; and the estimated time of departure (again a fictitious figure, such as 0000). These variables are labeled B, C, D, and E respectively.

The printer, having announced that it is about to do a Long Range Flight Plan, prints out the results for the overall trip immediately; it then waits for you to enter new speed, fuel, or time information for each leg, if you wish to, and to request results for the new leg by punching (2nd) (A′). If you don't enter any new data, the whole process takes about a minute. The printout shows what would come out for a trip from Tokyo to Los Angeles via Wake and Honolulu.

```
LONG RANGE FLT PLAN

   0.0000     WP
  35.4390     LAT
-140.1200     LON

   1.0000     WP
  19.1790     LAT
-166.3610     LON

   2.0000     WP
  21.1970     LAT
 158.0190     LON

   3.0000     WP
  33.5650     LAT
 118.2440     LON

 160.0000      GS
 350.0000    FUEL
   9.0000    BURN
   0.0000     ETD
4731.4890    DIST
  29.3418     ETE
   5.3418     ETA
 266.1463     EFR
  83.8537     EFL
```

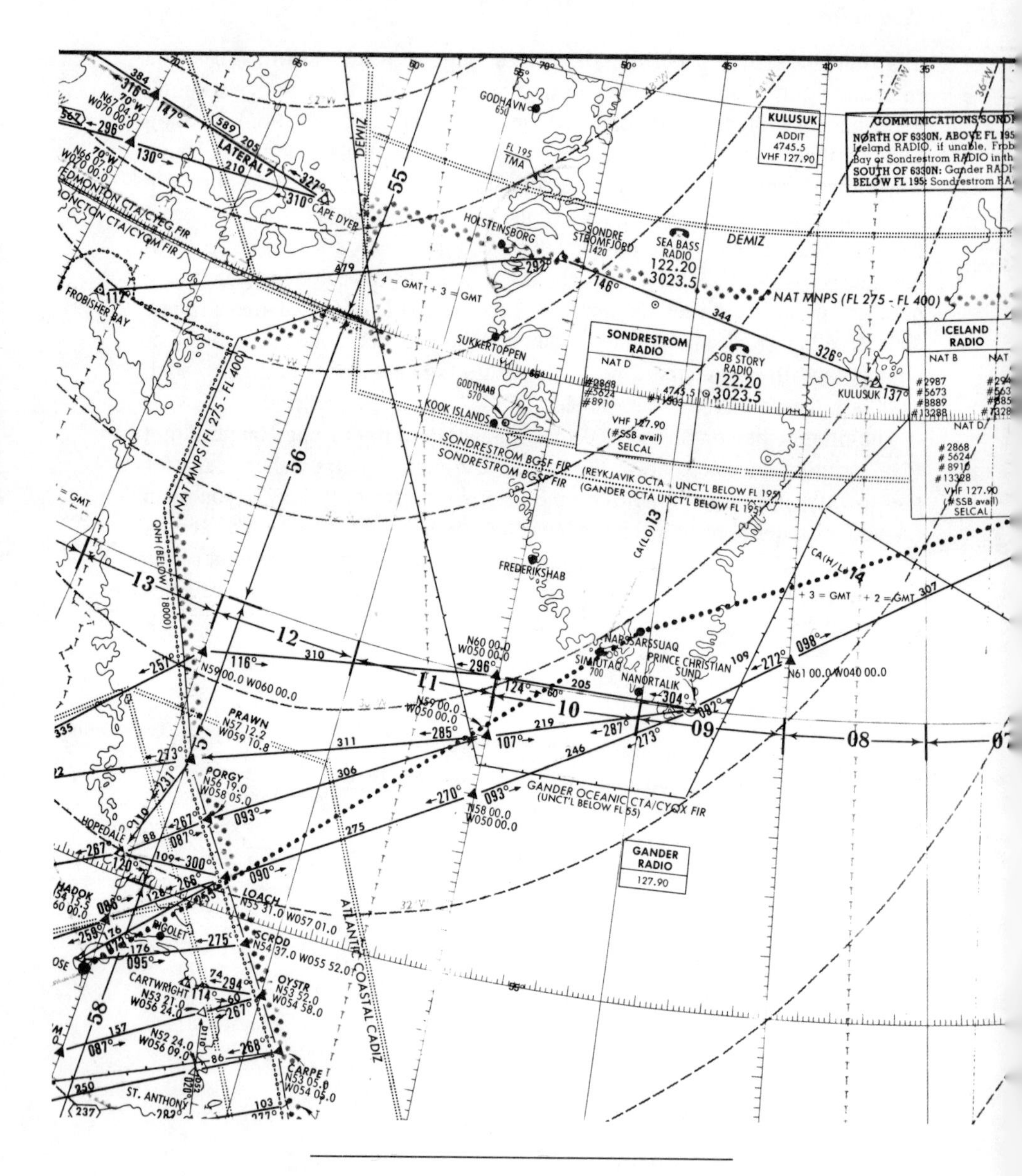

Plotting the route on the Jeppesen North Atlantic Orientation Chart.

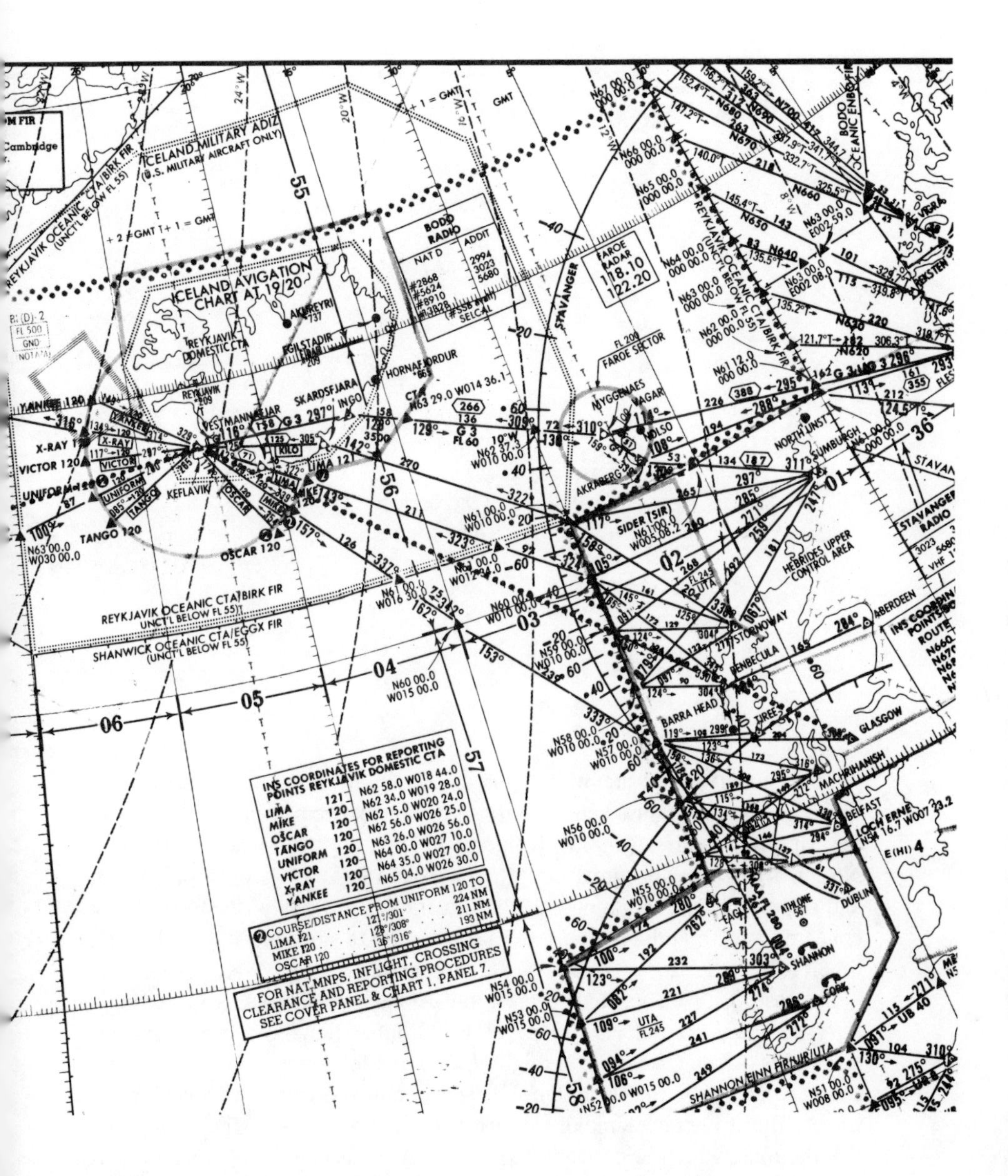

ICELAND MILITARY ADIZ
(U.S. MILITARY AIRCRAFT ONLY)
REYKJAVIK OCEANIC CTA/BIRK FIR
(UNCT'L BELOW FL 55)
ICELAND AVIGATION CHART AT 19/20
REYKJAVIK DOMESTIC CTA
AKUREYRI
EGILSTADIR
HORNAFJORDUR
SKARDSFJARA
VESTMANNAEJAR
KEFLAVIK
BODO RADIO
NAT D
ADDIT
#2868
#5624
#8910
#13828
2994
3023
5680
SELCAL
FAROE RADAR
118.10
122.20
FL 200
FAROE SECTOR
STAVANGER
MYGGENAES
VAGAR
AKRABERG
SIDER
NORTH UNST
SUMBURGH
HEBRIDES UPPER CONTROL AREA
STORNOWAY
BENBECULA
BARRA HEAD
TIREE
ABERDEEN
GLASGOW
MACHRIHANISH
BELFAST
LOCH ERNE
N54 16.7 W007 23.2
ATHLONE
SHANNON
CORK
DUBLIN
STAVANGER RADIO
SHANNON EINN FIR/UIR/UTA
REYKJAVIK OCEANIC CTA/BIRK FIR
UNCT'L BELOW FL 55
SHANWICK OCEANIC CTA/EGGX FIR
(UNCT'L BELOW FL 55)
N63 00.0
W030 00.0
N60 00.0
W015 00.0
INS COORDINATES FOR REPORTING POINTS REYKJAVIK DOMESTIC CTA
LIMA 121 N62 58.0 W018 44.0
MIKE 120 N62 34.0 W019 28.0
OSCAR 120 N62 15.0 W020 24.0
TANGO 120 N62 56.0 W026 25.0
UNIFORM 120 N63 26.0 W027 10.0
VICTOR 120 N64 00.0 W027 00.0
X-RAY 120 N64 35.0 W027 00.0
YANKEE 120 N65 04.0 W026 30.0
COURSE/DISTANCE FROM UNIFORM 120 TO
LIMA 121 121°/301° 224 NM
MIKE 120 128°/308° 211 NM
OSCAR 120 136°/316° 193 NM
FOR NAT, MNPS, INFLIGHT, CROSSING CLEARANCE AND REPORTING PROCEDURES SEE COVER PANEL & CHART 1, PANEL 7.

```
     1.0000      LEG
    19.1830     DLAT
  -166.3610     DLON
  1706.8954     DIST
  1706.8954     TDST
   118.2395      TC
    10.4005      ETE
    10.4005      ETA
    96.0129      EFR
   253.9871      EFL

     2.0000      LEG
    21.2010     DLAT
   158.0230     DLON
  1988.9300     DIST
  3695.8254     TDST
    80.2945      TC
    12.2551      ETE
    23.0556      ETA
   111.8773      EFR
   142.1098      EFL

     3.0000      LEG
    33.5650     DLAT
   118.2440     DLON
  2221.9840     DIST
  5917.8093     TDST
    61.4655      TC
    13.5315      ETE
    12.5911      ETA
   124.9866      EFR
    17.1232      EFL
```

The estimated total fuel of 350 gallons, assuming a 160 knot groundspeed and 9 gph burn, is just about right, since the fuel remaining at the end of the trip is 17 gallons (final EFL figure; EFL means "estimated fuel left"). Note that the nonstop great circle from Tokyo to L.A. is only 4,731 nm, and required only 266 gallons; this information is obviously irrelevant. What concerns you, mainly, is the total distance (5,918 nm) and the total flying time (10.4 + 12.25 + 13.53, or 36.18 hours).

Suppose that you wanted to cross the Atlantic in a C-182 with a single ferry tank and a no-reserve range of 1,300 nm. Allowing a two-hours' reserve plus a fudge factor, you might decide that the longest nonstop leg you can handle is 950 nm. Using the AV-04 program you could evaluate various routes. But it is much simpler to start with the globe, from which it appears that the route which best matches your limitations is Goose Bay (Newfoundland)—Julianehåb (Greenland)—Reykjavik (Iceland)—Glasgow (Scotland). The leg lengths on the globe appear to be 775, 790, and 830 statute miles respectively, or 674, 687, and 722 nm. These are all well within your capabilities, so much so that you might be tempted to look for a smaller number of longer legs. But there aren't any alternatives.

You might then turn to the Jeppesen Atlantic Orientation Charts, where you first discover that neither Julianehåb nor Reykjavik is charted. In their place are Narssarssuaq, Greenland, and Keflavik, Iceland. Measuring mileages, you get 667, 647, and 742 nm; the differences are due both to the different locations and to the errors of scaling on the conformal conic projection. From this chart you can also get initial headings, by taking a small, accurate protractor (from an art or drafting supply store, not a pilot's protractor) and placing its straight edge along the first meridian to the right of the point of departure, with the center mark on the course line. For the time being, you can ignore magnetic variation.

Turning now to the TI59, using lat/long information from the orientation chart, you can refine your information still further. In this case, you entered 110 gallons, the total fuel, at the beginning of each leg, and 0000 for the time of departure.

```
LONG RANGE FLT PLAN

      0.0000     WP
     53.1800     LAT
     60.2700     LON

      1.0000     WP
     61.1000     LAT
     45.2400     LON

      2.0000     WP
     63.5900     LAT
     22.3600     LON

      3.0000     WP
     55.4200     LAT
      4.2100     LON

    135.0000     GS
    110.0000     FUEL
     11.0000     BURN
      0.0000     ETD
   1906.1157     DIST
     14.0710     ETE
     14.0710     ETA
    155.3131     EFR
    -45.3131     EFL

    110.0000     FUEL
      0.0000     ETD
      1.0000     LEG
     61.1000     DLAT
     45.2400     DLON
    676.4405     DIST
    676.4405     TDST
     39.8317     TC
      5.0039     ETE
      5.0039     ETA
     55.1174     EFR
     54.8826     EFL
```

```
 110.0000    FUEL
   0.0000     ETD
   2.0000     LEG
  63.5860    DLAT
  22.3600    DLON
 648.4430    DIST
1324.8836    TDST
  65.0268      TC
   4.4812     ETE
   4.4812     ETA
  52.8361     EFR
  57.1639     EFL

 110.0000    FUEL
   0.0000     ETD
   3.0000     LEG
  55.4200    DLAT
   4.2100    DLON
 736.5664    DIST
2061.4500    TDST
 123.9018      TC
   5.2722     ETE
   5.2722     ETA
  60.0165     EFR
  49.9835     EFL
```

Note that the distances, 676, 648, and 737 nm, are in good agreement with those from the globe and the orientation chart; the headings, which are *initial* headings for great circle tracks, are almost precisely the same as those measured with the protractor. A fifty-gallon reserve remains at the end of each leg, which suggests several things. Either the ferry tank is unnecessarily big for the particular trip; or you could cruise faster; or you need have no fear of strong headwinds on the return; or, in a pinch, you could make the trip (though it would not be very wise) with the 182's normal long-range fuel capacity of 75 gallons, which provides a handbook range of about 900 nm at 55 percent power.

The TI59 has another program, AV-10, which can refine this flight plan still further. Since the legs are quite short, little will be lost by flying a rhumb line from waypoint to waypoint. The information is entered somewhat differently from that for the AV-04, and unlike AV-04 this program does not require the printer, and does not print out its results automatically. However, the answers look like this:

Leg 1 (Goose-Narssarssuaq): 678 nm, 46 degrees
Leg 2 (Narssarssuaq-Keflavik): 652 nm, 75 degrees
Leg 3 (Keflavik-Glasgow): 739 nm, 132 degrees

It is necessary to adjust these headings for the magnetic variation, which is 32 degrees west at Goose and 10 degrees west at Glasgow. With such large magnetic variations, it is important to keep careful track of them; the variation changes by 12 degrees on the way from Greenland to Iceland, and by 14 degrees on the way from Iceland to Scotland. But the results from AV-10 are the "correct" results for the trip; they are the ones on which your final flight plan will be based. In each case, the rhumb line heading is greater—that is, more southerly—than the initial great circle heading. But the rhumb line heading is a constant, which only needs to be adjusted for the magnetic variation. The increased distance over the series of rhumb lines is trivial—8 nm for the entire Atlantic crossing.

If you did not have the calculator to determine rhumb line courses, you would have to measure the heading at each meridian on the orientation chart, not just the first one, and then either average them out to get an approximate rhumb line, or plot a series of course changes at intervals (usually of 5 degrees longitude) along the route.

In this case, the waypoints were naturally provided by landing places along the route. Longer flights over open water do not offer these convenient waypoints, and you have to find your own.

Suppose, for instance, that you wanted to repeat the flight of Lindbergh, nonstop from New York to Paris, in a Mooney 201. Using the airplane handbook you find that you can expect to get 17 nmpg at a medium cruising speed of 155 knots. The distance on the National Geographic globe is 3,625 statute miles, or 3,152 nm. Because the last several hundred miles of the trip are over land, it will not be necessary to have as large a reserve as if the destination were also the first landfall; allowing a two-hour reserve at reduced power, you calculate that you will need 200 gallons of fuel. This means 136 gallons of ferry tankage, or about 18.2 cubic feet. As an aid in visualization, this would be a tank three feet long, three feet wide, and two feet deep. Assuming that you find a place in the airplane for all this fuel, and that you can manage the weight and balance, you now have but to prepare your route. Lindbergh used marine navigation charts, which he purchased at a ship chandler's; you will use the TI59, this time with program AV-11, Great Circle Flying. The purpose of this program is to pick the waypoints along the route; these will in turn be used to plot rhumb line courses.

Use the program to find the "vertices" of the great circle at each 10-degree interval. The results, again delivered in the twinkling of an eye, are:

	Longitude	Latitude	
NEW YORK	73.46	40.47	
	70	42.4645	42° 46′ 45″
	60	46.5036	
	50	49.3623	
	40	51.1809	
	30	52.0433	
	20	51.5930	
	10	51.0236	
	0	49.0901	
PARIS	1.59	48.39	

Note that while the orientation chart gives positions in degrees, minutes, and tenths of minutes, the TI59 is giving them in degrees, minutes, and seconds. No matter; everything may as well be rounded off to the nearest degree, because the oceanic controllers won't take flight plans or position reports otherwise than in whole degrees.

Making a series of dots on the orientation chart to indicate the calculated points, you find that the line takes you just south of Gander, Newfoundland, and right over the Landsend VOR at the southwestern tip of England. Since you will want to use both VORs, you alter the route to overfly both. You then forget about the portions of the flight between New York and Gander, and between Landsend VOR and Paris, because there you will be navigating by radio aids. The task is now reduced to plotting a series of rhumb lines connecting whole-degrees waypoints between Gander and Land's End. These are the waypoints.

	Waypoint	Longitude	Latitude
GANDER	1.	54° 32.2′	48° 54.0′
	2.	50	50
	3.	40	51
	4.	30	52
	5.	20	52
	6.	10	51
LANDSEND VOR	7.	5° 38.2′	50° 08.2′

Running these through the rhumb line program, you get the following:

Waypoint		Distance	True Course
1.		0	
2.		188.74	69.53
3.		386.32	81.07
4.		378.38	80.87
5.		369.40	90.00
6.		378.28	99.13
7.		174.35	107.35
	Total	1875.35	

This will be your complete flight planning information for the overwater segment. Since it is customary over the Atlantic for light aircraft to give position reports at 5 degree increments of longitude rather than ten degree increments, you can either fake it and give whole degree approximations of the latitude, or file whole degrees and then try giving position reports with half degrees and see if they are accepted. At any rate, with dead reckoning it is difficult to be certain of your position to within less than a degree.

If you run Gander–Land's End through the great circle program, you get a distance of 1,872 nm; so the penalty for the rhumb line approximation along the vertices of the great circle is only 3 nm. Obviously such small offsets from the course make very little difference. Interestingly, Lindbergh himself offset his course somewhat to the north, in order to have a landfall at Ireland. He aimed for the middle of the Irish coast, and after twenty hours over water, he hit it.

These examples show how information taken from a globe, referred to a planning chart, and then refined with a calculator can give precise flight-planning data. What is not obvious unless you have used the computer is the rapidity with which it produces its results. The entire process of planning the New York-to-Paris great circle, and getting waypoints, leg lengths, and true courses for the overwater segment, takes only ten or fifteen minutes once you have learned to use the keyboard. Furthermore, the information is probably more accurate than what you could derive from a planning chart, and certainly more accurate than your navigating could possibly be (given the inaccuracies of dead reckoning, the planning chart is quite satisfactorily accurate for light aircraft).

You ought to keep in mind, however, that a calculator is not necessarily so accurate as it looks. Computers have a way of blandly pumping out reams of numbers, each carrying a long tail of decimal places, that suggests fantastic accuracy. Actually, the computed information is not necessarily especially precise, and all the decimal places are mere artifacts of the workings of the machine. If you try computing a course back and forth a few times between two points using a dead-reckoning program (you insert lat/long for the start, and a heading, speed, and time, and it comes back with the new position) you will be disappointed and perhaps disillusioned to discover that the true positions soon get lost in a thicket of errors. What the computer excels in is speed and dependability; for the rest, a protractor and a pencil could do as well.

Having plotted a course, your next problem is to follow it. This is not always so easy as it seems during short flights over land. In IFR flying by VOR, there is never any inaccuracy because there is no dead reckoning. In VFR flying by pilotage, one has continuous opportunities to check the intended course against the landmarks as they pass by, and to make ad hoc corrections. Over water the conditions are quite different. For a period of ten hours or more there may be no opportunity for correcting the course. Errors are cumulative. One would like to think that they are random as well, but usually they are not. For instance, when you are hand-flying the airplane, you constantly find that you are wandering off heading; and usually you wander off in the same direction. With an autopilot with heading hold, the same problem occurs; but now it is due to the precession of the directional gyro rather than to coriolis effect or to some slight assymetry in the airplane or in the pilot's muscular equilibrium. As fatigue mounts the errors grow larger, both because one is slower to observe them, and because one feels more tolerant of them even when they have been caught.

For meteorological purposes the Atlantic is broken up into zones five degrees of longitude wide. Weather forecasting and reporting is very accurate, and the wind information is good—better than it is over land, often, because there are fewer disturbances. With a wind vector for each five-degree stretch and a precomputed heading, you can readily work out a corrected heading for each sector. You have then only to have swung your compass in advance, and to keep correcting the DG on a regular schedule if you are using it rather than the compass for heading reference.

An alternative technique is to sum the winds for the entire trip and fly a single correction. By doing this you allow yourself to be blown to the right or left of your course as the wind chooses, but you end up spot on at your destination if you have done everything perfectly. The calculator can be used to sum up a net wind. Suppose you have the following wind information (the numbers refer to five-degree sectors, counting westward from the Greenwich meridian):

11: 18015
10: 15012
09: 10005
08: 35008
07: 33015
06: 31025
05: 28025
04: 23020
03: 21015
02: 27005

That is, the wind in sector 11, from 55 to 50 degrees west, is from 180 degrees at 15 knots; in sector 10 from 150 degrees at 12; and so on. All these data are entered successively into the computer after calling up the wind vector program (AV-07); the final result is 260.38 degrees at 6.95 knots—in short, 26007. This amounts to a 7 knot tailwind; the cross-track component is negligible. Rather than bother with separate wind corrections every five degrees of longitude, the pilot can simply ignore the wind, and expect to arrive slightly early. He can correct his flight planned groundspeed by plus 7 knots for position reporting purposes, and then bias his reports slightly to compensate for the fact that, as he knows from the net report, he actually has a slight headwind for the first half of the trip, and picks up all his tailwind on the second half.

(Extreme accuracy in position reporting is neither possible, nor useful, nor expected. Since it is understood that small aircraft on ocean flights are probably using dead reckoning to navigate, the purpose of position reports is principally to verify that they are still airborne. In the event of a mishap, the search and rescue team would make use of the same dead reckoning procedures to begin its search that the pilot uses to report his position.)

A series of winds like these, viewed in conjunction with, say, a 700 mb (millibar) prog chart, might give an opportunity to try a little pressure-pattern navigation. This particular wind sequence suggests a low pressure area over Newfoundland, southwest of Gander, and another between Greenland and Iceland. There might be a high to the south of track, west of the Azores. Some of the strongest tailwinds will be coming from the left; the least-time track might deviate initially somewhat to the north, then turn somewhat more southward when the wind shifts, and follow the isobars, more or less, until the last sector, where it would turn straight toward Shannon and put the wind squarely on the tail.

The TI has a program (AV-02) for calculating heading correction (magnetic or true) and groundspeed with a given TAS and wind vector. The calculation can be performed, a little less conveniently, with any pocket calculator which does trigonometric functions. The formula for the heading correction is:

$$\arcsin\left[\frac{W}{V}\right] \times \sin(D-C)$$

where W is the wind velocity, V the airplane's TAS, D the wind direction, and C the true course. D–C is thus the angle of the wind to the true course.

Groundspeed is:

$$\left(V \times \cos \vartheta\right) - \left(W \times \cos(D-C)\right)$$

where ϑ is the drift correction angle computed in the foregoing equation.

This information can also be obtained using an E6B-type drift computer, or by drawing a wind triangle. In case you have forgotten how to draw a wind triangle, the procedure is:

1. Draw a line of indefinite length representing your true course.
2. Draw a line from the origin of the first line, representing the wind direction (always true, not magnetic).
3. Measure off along the wind line a length representing the strength of the wind. To keep the drawing to a convenient scale, use 1/4 inch = 10 knots, or 1/2 inch = 10 knots.
4. From the free end of the wind vector line draw a line to the true course line, of a length corresponding to your TAS.

The distance from the origin to the intersection along the true course line represents your groundspeed. The angle of the diagonal line from the end of the wind vector line to the true course line represents your true heading.

Rather than orient the true course line to match the true course on the map, you can simply align it vertically or horizontally and draw the wind vector at the appropriate angle to it. The correction angle will then be represented by the angle of the third line to the first.

A little practice in drawing wind triangles is recommended. It will clarify, among other things, why the majority of winds are headwinds, even those blowing from directly abeam.

Alterations to preplanned courses may be made en route. In daytime, you can judge the wind direction and strength from the appearance of the ocean surface, and compare what you observe with the forecast. In light airs, the surface of the water is streaked with lines called wind lines. By watching

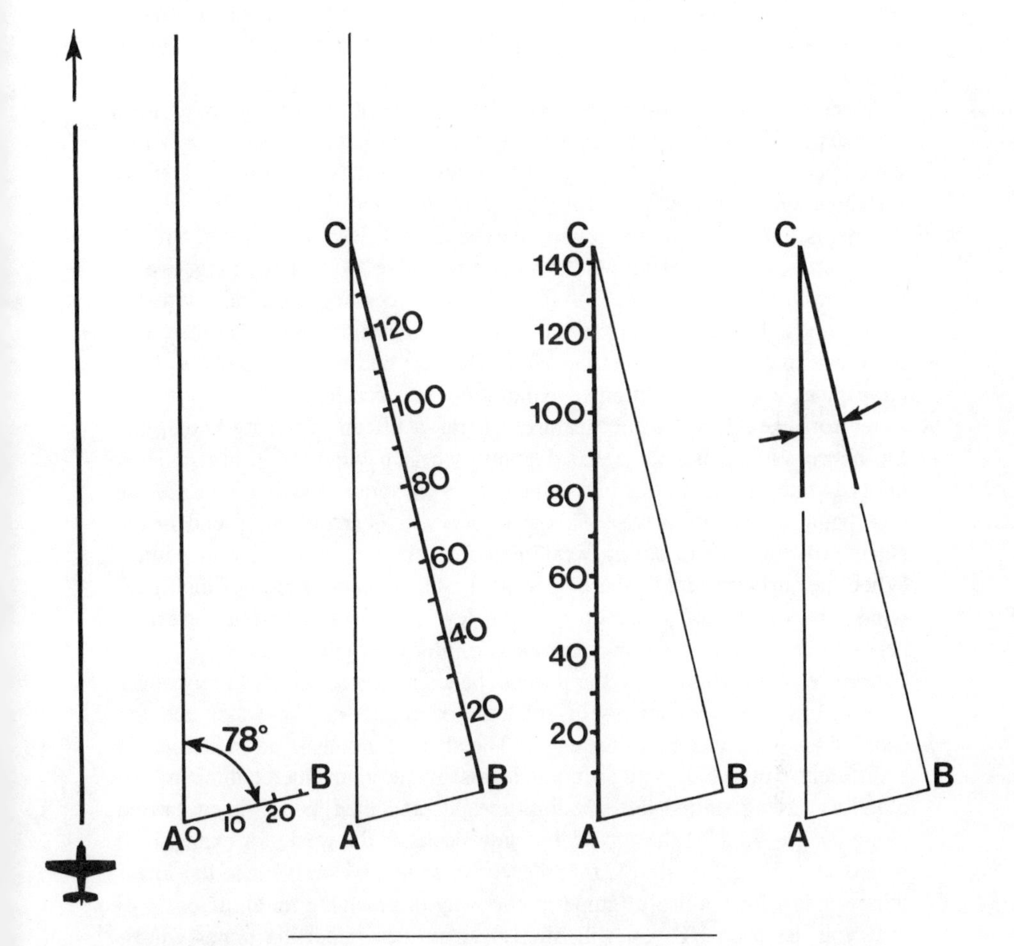

The first line represents your desired track on any heading. You scale off a wind vector, here 25 knots at 78 degrees to the desired track, at one end. Then you scale a line representing your TAS (here 135 knots) from the end of the wind vector back to the track line. AC now represents the ground-speed made good—142 knots—and angle ACB is the wind-drift correction, about 9 degrees. Knots, mph, and magnetic and true directions offer many opportunities for errors; fortunately life offers few opportunities to use wind triangles. The TI59's answers to this problem, incidentally, are 142.31 knots and 9.89 degrees drift correction.

them carefully, you can judge from the way in which they propagate over the water which way the wind is blowing at the surface. Light surface winds, however, are not necessarily indicative of the direction of upper winds.

When wind speed increases above ten knots, the water begins to form whitecaps. The direction of the wind is not immediately evident; the whitecaps appear to slide off the ridges of waves, and it is tempting to identify the direction of their apparent motion with the wind direction. In fact, the reverse is true: the wind is blowing the waves out from under the foam.

Sometimes several patterns of waves are visible. The large, long swells, with amplitudes of hundreds of feet, do not necessarily coincide with the local winds. It is the choppy swell, with an amplitude of several feet, that is blown up by the local surface wind. The size and turbulence of the local chop gives a good indication of wind strength; occasional whitecaps occur at ten to fifteen knots; whitecaps are general at fifteen to twenty knots, and the entire surface has an agitated appearance. In winds of thirty to forty knots, whitecaps blow out into long streaks of spray, and one can see the tops being blown off waves. This kind of water is frequently found in the North Atlantic, where strong westerlies are common, especially in winter. When the surface wind is a strong wind, it is a good indication of the upper wind. The winds-aloft reports for several levels will suggest what correction needs to be applied to the surface wind to infer the upper wind.

Surface observations give the pilot a chance to correct his dead reckoning information. The movements of shadows of clouds on the water and the slant of towering clouds do the same. The slant of clouds is an indication of a difference in velocity at different levels in the atmosphere; think of the cloud as having started out standing upright, and then having been carried along by the wind. If the top of the cloud leans to the west, for example, it means that the upper wind is more easterly, or less westerly, than the lower wind—sometimes a useful guide in choosing or changing an altitude.

If you listen on HF you will often hear airliners reporting upper winds. They are not necessarily of much use to light aircraft, since stratospheric winds may differ greatly in direction and force from lower winds. For instance, while bucking a fifty-knot headwind over the tip of Greenland, I heard a jet reporting light and variable winds at 41,000 feet.

During daytime, and under some conditions at night, an airliner can provide a position fix, if you happen to catch sight of it or its contrail. The matter is particularly facilitated if you can identify the aircraft type. Using 121.5 kHz, the emergency frequency which all ocean flights monitor, you make a blind call to the DC-10, 747, or simply the airliner (it could be an Ilyushin, for that matter) eastbound at such and such a latitude and longitude. If you establish radio contact, you go over to 123.45 and request that they do something to identify themselves, such as make a slight turn or, at twilight or at night, flash their position lights or landing lights. Once you have a definite identification, you can get their Inertial lat/long readout to update your own dead reckoning calculations.

At night, it is difficult to use any of the empirical means of judging wind drift. However, it may be possible to make use of a sextant, if you know how to use one, to get a rough idea of your position. A sun sight can be taken in daytime, for that matter. Because of the visual distortions caused by the windscreens and the limited number of celestial bodies which will present themselves in a good location for taking a sight, celestial navigation has been used very little by aircraft since before World War II, and least of all by light aircraft. However, since great accuracy is not required, and any position within twenty or even fifty miles of the true one is better than no position at all, some ferry pilots do occasionally use celestial. An accurate chronometer is required, but with electronic watches that no longer poses a problem. At the very least, a sextant might help while away the time on a long flight, even if it could not be depended upon for precise navigation.

In addition to being able to make dead reckoning judgments from observations of surface winds and clouds, ferry pilots speak of having a *feeling* of being on one side of the course or another. While it would be difficult to make objective tests of the validity of this feeling, it is certainly true that one often has some kind of hunch about being right or left of course, and on many occasions this hunch may turn out to be correct. On the other hand, it is not reliable, and pilots depend on it only when they have reason to believe they are not where they should be—when, for example, they do not make a landfall when they should—and have to make a decision, in the complete absence of objective information, about which way to turn.

One ferry pilot, speaking of the accuracy with which it is possible to dead reckon over long distances, told of getting a position fix from C-141 above him mid-Pacific. The C-141—a four-engine Air Force jet transport—was relying on Doppler radar for its navigation; this system, which is obsolescent, involves reading speed and drift across the earth's surface by means of a radar signal, and integrating this information continuously, as Inertial does, into an updated position. The ferry pilot disagreed with the C-141's report of its position, and when the pilot later spoke with an officer who had been aboard the jet, he learned that the navigator had rechecked the position and discovered that he had made an error, and indeed the ferry pilot's dead-reckoned position had been the more accurate.

Nevertheless, because weather conditions do not always permit observations of the surface and forecast winds can be significantly in error, dead reckoning is not always altogether dependable—despite its occasional shining successes. Louise Sacchi, in her book, *Ocean Flying,* absolutely discourages private pilots from attempting the 2,000 mile Gander to Shannon flight, or that from California to Hawaii, without some kind of long-range navigation radio—namely Loran. My own experience, flying a fairly fast airplane over ocean legs ranging in length from 900 to nearly 2,000 miles, has been of one major miss in five flights. The miss—I was off-track by 135 nm on an overwater leg of 1,300 nm—was not accompanied by any indication or premonition of error and could have been disastrous under other circumstances, for instance if the angular error had been extrapolated to a flight of 2,000 nm or more, and if an ADF had failed.

The fact of the matter is, however, that most ferry flights across the Atlantic do go directly from Gander to Shannon, and they do so without Loran; and the same is true of many Pacific flights from the mainland to Hawaii. So casual is the general attitude toward Atlantic navigation, in fact, that one ferry-firm operator told me that all you have to do is set your wet compass on E—for "Europe"—and you'll get there eventually. This seems an excessively approximate heading, since the actual headings across the pond are from 95 to 112 degrees, but it seemed reassuring at the time.

Some ferry pilots flying the California–Hawaii route mention a mysterious 5-degrees-left (or more) course correction which must be made to avoid ending up north of the islands. Sacchi does not. The reason for the correction is obscure; possible explanations include gyro drift and uncharted low-level magnetic variations, systematic errors in wind forecasting, or what you will. Coriolis effect has been mentioned; this, however, would make the airplane tend to turn to the right (to the left in the southern hemisphere), but would not affect its compass. To the extent that a cumulative error would result if every time the airplane wandered off its heading it did so to the right, the Coriolis effect might play a role.

An instructive exercise for a pilot planning a long ocean flight is to dead-reckon over land, preferably at night, for a long distance—over 500 nm—and to observe the magnitude of his errors. The longer the flight, the better; and of course it is important not to allow stray information about position to enter one's mind (hence the preferability of a night flight). A little overland practice will make clear the scale of the likely errors, the nature of the routines used in planning legs and headings (which are exactly those used in private-pilot training to develop wind triangles), and the difficulty that will be experienced in holding heading, staying awake, and so on. Dead reckoning over land is made easier by the ready availability of low-level wind information, but there are still plenty of errors to throw you off.

Barry Schiff, airline pilot and aviation writer, once suggested in *Air Progress* magazine that people contemplating ocean flights try aiming straight out to sea, flying about 500 miles or so, and then turning around and coming back. This was just for practice, mind you. I read this suggestion and decided that it was reckless, because whereas on an ocean crossing you are going somewhere and just happen to be passing over the ocean, if you poked yourself way out there and then turned around and came back it would be an act of provocation. It would be asking for trouble. So I never did it; it seemed to be a much more frightening idea than that of jumping across from this side to that. However, everyone may not share my superstitious dread of Poseidon, and it would certainly be good practice. It would be extremely instructive to see where you made your landfall on the return flight, and to keep a log of sea surface observations and of your impressions about drift, and later to compare them with the actual drift you experienced. And if such a flight seems, when you think about it, a little like going out on a limb for nothing—or even like walking the plank—it might give you some idea of how you will feel when you find yourself in the middle of the ocean and, no matter how diligently you've been navigating, you realize that you really don't know where the hell you are.

5

METEOROLOGY

Long flights encounter weather on a different scale than short ones do. There are analogies between the two scales; just as a pilot on a 200-mile flight might take a detour to avoid an area of cloud buildups over a ridge between him and his destination, a pilot on a 1,000-mile flight will detour around a front with embedded thunderstorms, or swing to one side or another of an entire weather system. The difference is in the approach to the problems. In one case, the pilot sees the clouds ahead of him, locates a clear area or a low spot where he thinks he can get across, and heads for it. The pilot on the more ambitious flight cannot always afford to wait until he sees the obstacles in his way. If they take the form of wind, the information that reaches him in the form of drift and groundspeed is incomprehensible unless he has taken in the larger weather picture; and if, as he motors along in darkness, he gradually becomes aware that the entire horizon ahead of him is flickering with thunderstorms, it is already too late for him to make a deviation—or at least to make one that is economical of time and effort.

In overwater flying, weather presents two principal elements of hazard: wind and icing. The former, if it is not taken into account, can drive an airplane far off course, make it miss an island destination, or cause it to run short of fuel; and because of the uncertainties of dead reckoning over the ocean, particularly in bad weather, the pilot would have no way whatever of knowing what was happening to him.

Icing over water is in a sense less serious than over land, because there are no obstacles (except icebergs) and one could in a pinch descend to just above the surface, where the air would presumably be warmer and the engine would develop its maximum power. I have heard it said that icing is not as risky over the North Atlantic as elsewhere because the waters of the North Atlantic are comparatively warm. The Gulf Stream flows along the entire Newfoundland–Ireland route (but not up to Greenland and Iceland), and so the freezing level is scarcely ever right down on the surface. I am not certain that this is true, but it is a thesis that ships have no doubt exhaustively tested. If you collect a heavy load of ice and cannot remain airborne, you could splash even though there was a layer of warmer air near the surface, because there would not be time during your descent for the warmer air to melt the ice away. In the Pacific, the northern route to the Orient via the Aleutians is little used by ferry pilots because the risk of icing is severe all the way from the Alaska panhandle over to Japan during much of the year; the central and southern Pacific routes, on the other hand, are always free of ice.

In overland flying, wind is not usually a matter of life or death, because there are almost always alternate airports, or in a pinch roads or fields, on which to land if for some freakish reason a pilot runs out of fuel short of his destination or is blown far off course. There are very few places where this could happen, unless the pilot were sleeping and his alarm clock failed to ring, since the sprinkling of navigational aids over most land areas makes major navigational errors unlikely. Perhaps in extreme northern Canada in IFR conditions a flier might have the same troubles with wind that he can easily have over the ocean.

Icing, on the other hand, is a worse problem over land than over water, because of the higher minimum altitudes and the much greater likelihood of the freezing level being at the surface. Extreme turbulence is also more probable, and thunderstorms are generally of greater intensity. (In fact, the American midwest brews some of the worst thunderstorms to be found anywhere. By contrast, in Europe a thunderstorm is not considered an absolute obstacle to an airplane, as it is here.)

The long-distance pilot is taking on weather, one could say, on the global scale. He is concerned with large weather phenomena. Not that small ones don't affect him as well; he in fact must cope with both the small local conditions and the big ones extending over hundreds and thousands of miles. Unlike the short-distance pilot, however, he is dealing with entire weather systems. He can no longer think in terms of the "flying conditions," because they will vary so much for him. The conditions at his departure may be of one sort, those en route of another, and at his destination of yet another.

On a long flight, when you are preoccupied with range, efficiency, speed, and with minimizing your own fatigue, you want not merely to accept the weather as it comes along on a straight line between your departure point and destination, but to make the most of it; to bend the line in such a way as to profit from the weather. As a hiker does not make a beeline from one place to another, up and downhill, but follows contour lines and picks the easy terrain to cross, a pilot can do the same. Weather itself forms a terrain, with uphill and downhill (upwind and downwind) sections, rough areas, natural obstacles, and so on. The problem for the pilot is to interpret that terrain.

Most people have a highly simplified idea of the structure of weather. To a person who never thinks about weather at all, it seems to materialize out of thin air, and subsequently to dematerialize; statements that a storm is "moving down from the northwest" seem metaphorical, like the statement "gloom descended upon the gathering." A more sophisticated observer realizes that the weather that he experiences generally arrives from somewhere else, unless he lives right next to some weather-producing feature, like a mountain range. His view takes in the role played by air masses—bodies of warmer or colder, drier or wetter air moving about the globe—in generating clouds and precipitation. A person who looks at the daily weather map in the newspapers would appreciate a few other things, for instance the fact that wind and weather are not the same thing, and that the direction in which the wind is blowing is not necessarily the direction in which the weather is moving. It is apparent even from simplified weather maps that there is some tendency for wind flows to parallel isobars, especially when the isobars are close together, and also that there is some connection between low pressure areas and wet weather, and between high pressure areas and fair weather—as the legend on a barometer would suggest. Rising barometer, fair weather ahead; falling barometer, storm coming.

This brings us to the level of meteorological sophistication expected of even the most casual pilot. But it does not tell us much that is of use for long-range flight planning.

Although much is made by some pilots of the value of television and newspaper weather presentations, as opposed to the notoriously undependable aviation weather forecasts obtained from Flight Service Stations, their value for en route planning is limited, because they are principally surface forecasts. People, after all, are mostly found on the surface. The first concept that a pilot has to get through his head is that air masses do not possess vertical boundaries, like so many walls and fences. Instead, they resemble a hilly landscape, with the complication that there may be levels of hills of different densities piled upon each other. Picture one of those conversation pieces that contain blobs of oil of different densities and colors which do not mix, but slide about upon one another. Then visualize the oils in a fairly torpid state (rather than the volcanic one that the manufacturers like to produce by putting a heat source at the bottom of the vessel); now you begin to have a picture of air masses in the atmosphere.

Surface charts do not provide a satisfactory picture of flying weather because they do not reveal the positions and relationships of systems above the surface; and these are the ones through which you will fly. This is particularly true in respect to wind. Whether you encounter clouds here or there does not greatly matter; but the way the wind is blowing does.

Weather charts are conventionally made for a number of levels, starting at the surface and running up through 850, 700, 500 millibars, and so on. The millibar is the metric unit by which atmospheric pressure is measured; the surface standard pressure of 1,013 mb corresponds to the value of 29.92 inches Hg (inches of mercury) used in the United States. The 850 mb level is at about 5,000 feet; 700 is at 10,000; 500 at 18,000. These are the altitudes that concern the light-aircraft pilot; the next level, 300 mb, is out of reach.

The surface chart consists of pressure contours, corrected to sea level and expressed either in millibars or inches of mercury. The contour lines are called isobars (from Greek roots meaning "same depth") and connect points of equal barometric pressure. The interval between isobars will differ, depending on where you find the map; but on a detailed map it may be either 4 mb or .04 inches Hg. Reporting stations are represented by a coded circle resembling an apple pierced by an arrow. The direction and number of feathers on the arrow indicate the surface wind. The surface wind is often unrelated to the orientation of the isobars, because surface wind is strongly influenced by terrain effects.

The upper level charts look similar, but are actually somewhat different. For example, the 500 mb chart will also show contour lines, which are commonly called isobars but are in fact "iso-heights." The upper level charts are contour maps of a "surface." The 500 mb surface, for example, is the height in the atmosphere, in meters above sea level, at which the absolute pressure is 500 mb. The iso-heights will be labeled (for example) 570, 564, 558, and so on; these figures mean that along that line the 500-mb surface is 5,700 meters above sea level, or 5,640 meters, etc. Iso-heights are read in the same way as isobars; lower values of pressure surface height mean lower local pressure. On the upper level charts, wind directions may or may not be indicated; if they are, the arrow with its feathers is still there, minus its apple, and the arrows will be seen to align themselves more or less with the contour lines, because the upper level winds obey pressure systems without disturbances from surface friction.

Because of an intricate balance of forces, of which the Coriolis effect is the principal counterpoise to the attraction and repulsion of lows and highs, winds blow in quasi-orbital paths, not directly toward lows or away from highs, but in spirals around them. The wind is like a car or cyclist racing around a banked track; the speed increases with the angle of bank, and the bank of the track corresponds, in the case of the wind, to the pressure gradient. A steep pressure gradient is represented on the weather map by isobars clustered closely together.

The atmosphere behaves somewhat like the curved space of Einsteinian physics. Just as objects in space fall toward "low spots," that is, toward gravitational fields, the winds fall toward, and more or less orbit around, low centers. Conversely, they spiral downhill away from highs. From the point of view of the pilot, the trick is to make the best use of this rolling terrain—that is, to fly downhill as much as possible.

Apart from considerations of weather—cloudiness, rain, ice, and so on—the topography of highs and lows is useful in selecting the route that will be most efficient and rapid. In an airplane, distance flown over the ground does not necessarily concern us; it is time en route that determines fuel consumption, fatigue, and self-satisfaction. Because of the wind, it is quite possible to go a longer distance in less time, and certainly equally possible, by making all the wrong choices, to do so in *more* time.

In the Northern Hemisphere, circulation around a low is counterclockwise, as seen from above, and that around a high is clockwise. Circulation around a low is called "cyclonic," and that around a high, "anticyclonic." Satellite pictures of storms often show a pinwheel structure in which the motion of the wind in toward the center and counterclockwise around the center is quite evident.

When the wind is behind you in the Northern Hemisphere, a low center would be to your left, and a high center to your right; so in your route planning you would like to keep things that way at all times. Unhappily, it is not always possible; in fact, it is rare to find pressure patterns cooperating precisely with your wishes. But it is not uncommon to find a low or a high somewhere along your route, and to be able to take advantage of it.

Though we habitually think in terms of straight lines and deplore doglegs as wasteful of time, it is actually possible to make a fairly wide dogleg without increasing trip time by very much. Suppose that on a 1,000-mile flight we have to deviate 100 miles to one side of a straight line. If the maximum deviation is placed at the center of the distance, say to the right, then we would fly about 11 degrees to the right of the direct course for 500 miles, and then make a left turn of 22 degrees and proceed to the destination. The extra distance traveled will be only about 20 miles. The closer to the origin or destination the point of maximum deviation is placed, the greater the extra distance becomes; if, for instance, we fly toward a point

100 miles to one side of the destination, and upon reaching it make a 90-degree turn toward the destination, then the extra distance traveled will have been a little more than 100 miles, or 10 percent. In order to make such a dogleg at least as rapid as the straight line, it would be necessary for it to produce a 10 percent improvement in overall speed; whereas in the first example, where the deviation was at the center of the course, only a 2 percent gain in speed would be needed.

Applied to an actual situation, this means that if we choose to follow a sinuous course which roughly parallels the isobars, rather than a straight line cutting across them, the tailwind contribution required to make it worthwhile must be greater the closer the point of the dogleg is located to either end of the course.

Following isobars in order to have the wind at your back is called "pressure-pattern navigation," and the optimized pressure-pattern route is called the "least-time track." There may be rigorous analytic methods of using this system, but I don't know what they are, and I doubt they would be of much interest to lightplane pilots. What is obviously useful, however, when long distances are involved and the scale of a flight is such as to take it across whole systems of highs and lows, is to understand the principle and to apply it when it seems advantageous to do so. Looking at the weather map as a contour map of the aerial "terrain" over which you will fly, you first locate your points of origin and destination. Then you locate the high and low centers adjacent to your track. The wind arrows scattered along your course assist you in getting a sense of the overall circulation. You then look for areas in which the isobars cluster together comparatively tightly and assess the orientation of these areas with respect to your course. If there is a stretch of favorable and steep gradient located close to your course and at a shallow angle to it, you consider the possibility of bending your track so as to fly to the area of steep pressure gradient and then to steer parallel to its isobars or slightly toward the low center (or away from the high center) with respect to the isobars. (Usually the wind flows at a slight angle to the isobars, toward the dominating low or away from the high, this angle being greater closer to the surface.)

On a trip like that from Gander to Shannon, where no considerations of the location of navaids or natural obstacles exist at all, and the route is laid out solely in terms of lat/long points, you may as well pick your route to match the general flow of the isobars, if there is a general flow, depending simply upon "feel" and your eyeball to judge how big an excursion from a straight line (or great circle) is permissible for how strong a compensating tailwind component. Remember that even if you cross isobars diagonally, you will still experience some tailwind or headwind component. The magnitude of the component can be determined by using a drift-angle computer,

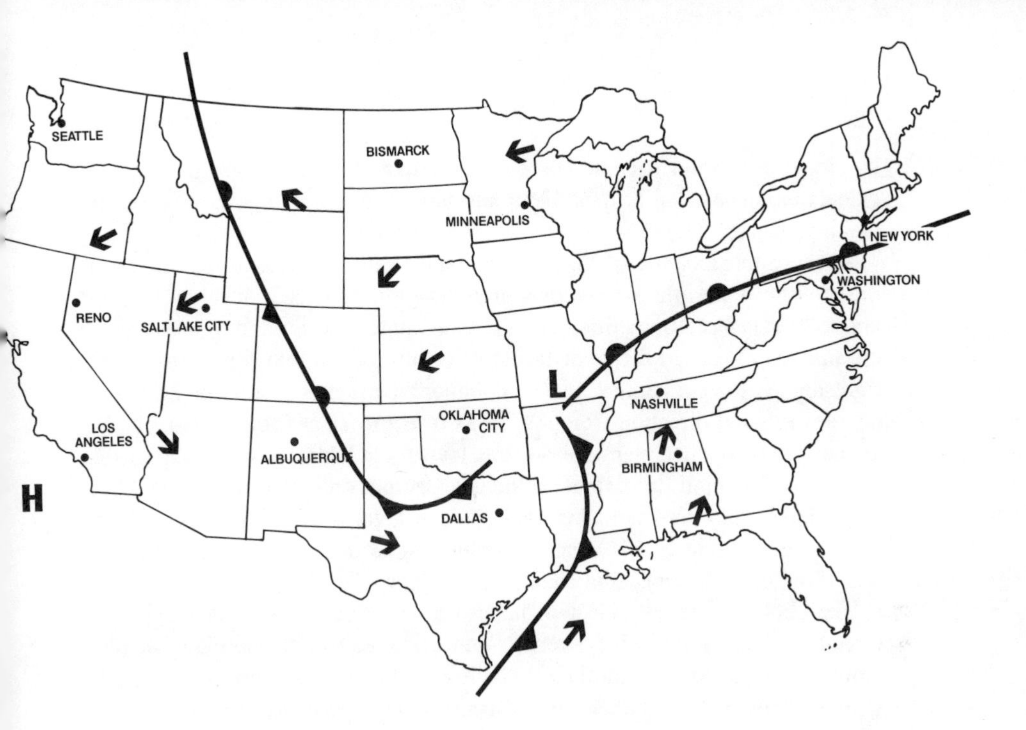

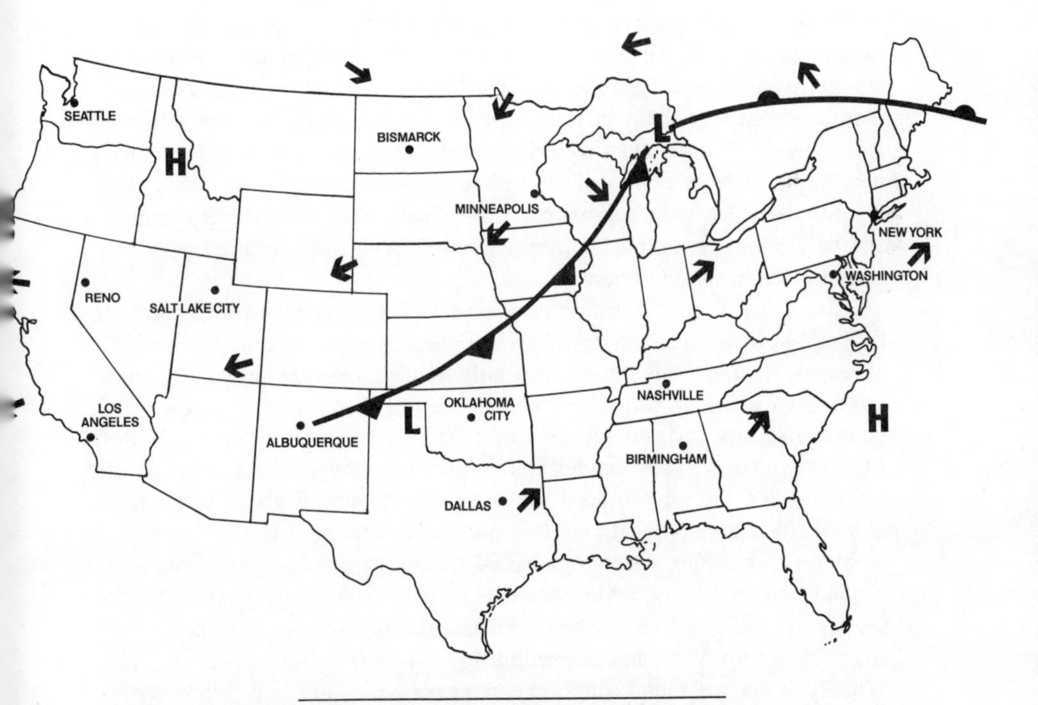

Two situations taken from newspaper weather maps. On November 9, going from Albuquerque to Nashville, you would have flown via Dallas, not Oklahoma City: an example of using a dogleg to capitalize on circulation. On December 11, going from Bismarck to Reno, you might have flown south of Salt Lake City. Note the concentration of precipitation to the north and east of the low centers.

by drawing a wind triangle, or by calculation, preferably using a pocket calculator. (The wind drift problem and its solution were discussed in more detail in Chapter 4.)

Because the location of highs, lows, and fronts varies at different levels in the atmosphere, the wind blows in different directions at different heights above the surface. Sometimes the effect is quite drastic, and you can look up and see clouds at different levels traveling in opposite directions, or at right angles to one another. More commonly, there is a shift of 20 or 30 degrees in wind direction from, say, 3,000 feet to 12,000 or 18,000. If you do not have weather maps for several levels available, you can still ask a briefer for the winds at several altitudes to see not only how strong they are, but whether they are more favorable at one level than another. A shift of only 10 degrees in a cross-track or quartering wind, if the wind is strong, will show up as a noticeable change in groundspeed.

The choice of an altitude will therefore be influenced not only by wind direction, but by the wind strength and the speed of the airplane at that altitude; a turbocharged airplane, which goes faster and faster the higher it climbs, may make a better groundspeed at high altitude even though an adverse wind there is stronger than it is at a lower level.

The large-scale wind systems are not the only ones which interest the long-distance pilot. Air flows in swirls and eddies like water in a stream or smoke rising from a cigarette, and the local circulation may be quite a bit more complex than that which is suggested by large highs and lows. Winds-aloft reports must be taken with a grain of salt, because they are developed by computers from bodies of data of sometimes uncertain accuracy, using algorithms which only approximate the actual behavior of the atmosphere; and they represent a probable average value over a large area during a long period of time, usually twelve hours.

Using DME, you can determine your actual wind component en route. It is common to see fluctuations of several knots in groundspeed over a period of several minutes; sometimes, over hilly terrain, there are quite large spurts and lags in speed. If you have a general idea of the position of a low along your course, use DME to optimize your flight around it; or if the winds-aloft forecast indicates a tailwind during the latter part of your flight, you can use DME to spot the groundspeed increase and to begin a climb to a higher altitude. When the wind forecast is uncertain, you can use DME to select an altitude, watching your groundspeed during the climb. A turbocharger is a good companion for DME, because it makes a large range of altitudes readily available; if you get a hunch that you might make a better groundspeed higher up, you can easily climb 2,000 or 4,000 feet and check it out. When you make a round trip over a short period of time, say within a day, you can make a mental note of the wind conditions on the trip out, and possibly adjust your return route accordingly.

Occasionally, local cloud formations indicate areas where better groundspeeds can be made. Flying below an area of fair-weather cumulus clouds, you sometimes see them arranged in lines, called cloud streets, parallel to the wind direction. If you are flying in the same direction, you would want to fly beneath the clouds, not in the gaps between rows, because the clouds cap columns and areas of rising air. In the vicinity of a mountain ridge across which the wind is blowing, you would like to fly along the upwind side, again to take advantage of rising air. Flying directly over the top of the ridge is useless, since you merely get a strong crosswind. Well downwind of a ridge, there are sometimes wave clouds that flag the locations of rising and falling streams of air. Some wave clouds are oval or lenticular in shape and aligned with the wind; others resemble elongated, flat loaves, with their long axes parallel to the ridge. The lenticular cloud stands still, and air rises into it at the edge near the ridge that triggered the wave, and sinks out the other edge. If you fly parallel to such a formation, you would always want to fly upwind of the edge of the cloud nearer the mountain.

It is not usually possible to make a great deal of use of local airflow phenomena, except in freakish cases; but in the interest of general efficiency, and of keeping in control of all the details of your flying, it is desirable to have an awareness of air as a fluid in motion, like a river, and to try to make the most of its interactions with the terrain. So far as the larger weather phenomena are concerned, an awareness of them is not merely desirable; it is indispensable.

When it comes to figuring out your course in detail, the big problem is to know where the weather is really going to be when you get there. A good weather analyst—places like Gander have some excellent ones—can give a reliable prognosis based on the movement of the pressure patterns over the last day or so. At the very least, you would want to have a look at a couple of charts of different ages yourself to assess the movement of systems. Over oceans, the movements of weather are more predictable than over land, because there are fewer surface features to influence them.

If a high lies squarely athwart your track, for instance, and is moving very slowly, and the gradients around it are shallow, you would plan to fly straight across it. A low in the same position, especially if its gradients were steep, would suggest that you should swing around to the right of it. A low moving across your track from the left would incline you both to swing to the right and if possible to take off sooner rather than later.

The amount of weather produced by a low depends on many factors: moisture content of the air, vertical instability, pressure gradients, surrounding pressure patterns. The structures of air masses and fronts become hopelessly interlaced and analysts become hopelessly confused. There is, however, a classical pattern which underlies all weather analysis, and which pilots have to understand.

Most temperate zone cloudiness and precipitation occurs because of the existence of fronts. Fronts are boundary zones between air masses of different temperatures. We usually think of fronts as being produced by low pressure areas, because most of the significant frontal weather we deal with is associated with a low; but it is possible for a front to exist in an area without a significant pressure gradient (a "stationary front"), and it is possible for the cart to get ahead of the horse, and for air masses, as a consequence of some random (that is, unforeseen) perturbation, to trigger the formation of a low pressure area. For the purposes of understanding flying weather, however, it is generally sufficient to assume that fronts occur in the vicinity of a low, but not of a high, because the effect of a low is to pull different air masses into contact, whereas that of a high is to push them apart. When you consider that a low consists of a spiral of air, like a shallow whirlpool, hundreds of miles across, and that there is a definite temperature gradient from pole to equator, it is apparent that warm air from the south will be pulled poleward and eastward by the cyclonic rotation, and cold air will move in the direction of the equator.

In the classical picture of a Northern Hemisphere low and its associated fronts, a broad wedge of warm air to the south of the low center is sandwiched between two masses of cold air. The one behind it to the north and west is sliding underneath it along the surface of the earth, wedging the warm air upward; the warm air is in turn sliding up over the cold air ahead of it, to the north and east. Underrunning cold air produces a cold front; overrunning warm air, a warm front.

In order for the classical frontal development to occur, air masses of the standard type have to be available. When they are not, frontal development will follow different patterns. On the west coast of the United States, for instance, there is very little warm front weather, because there is not a supply of warm moist air to the south to be pulled up over local cooler air by a low. Instead, cold moist air from the northern Pacific moves in under a large area of relatively warm dry air dominated by an offshore high, and frontal systems consist only of a cold front, with perhaps only a vestigial warm front close to the low center. Long periods of low ceilings, drizzle, and freezing rain rarely occur.

Cold front weather differs recognizably from warm front weather. For one thing, the slope of a cold front is steeper than that of a warm front, ranging, say, from 50:1 (50 miles horizontally per mile vertically) to 100:1. The weather along the front covers a relatively narrow band of considerable vertical depth. The weather may be quite violent, with a strong wind shear across the front, but after frontal passage, clearing is rapid. Typical postfrontal weather consists of scattered clouds below 10,000 feet under a clear sky with lower temperatures and strong, gusty winds.

Warm front weather, on the other hand, occurs along a flatter slope, with a gradient of 100:1 to 300:1 (meaning that the frontal surface at an altitude of 10,000 feet may be 300 miles ahead of the frontal surface at 5,000 feet). It is characterized by large expanses of stable, layered clouds, long periods of steady precipitation, and often by dangerous icing conditions from which the only escape is to climb up into the overrunning warm air—a difficult trick when an airplane has already collected ice.

The cloudiness associated with a warm front classically extends from the southeastern quadrant of the low around to the northeastern quadrant, as moist, warm southern air is lifted and drawn northward at the same time. It is a rule of thumb about low pressure systems that the largest area of the most unpleasant weather will be found in the northeastern quadrant of the low.

The positions of fronts are marked on isobaric charts by inflections—sharp bends—in the isobars; the bend is invariably in toward the low center, and consequently the wind bears more from the right after you fly through a front, regardless of your direction of flight. (The opposite rule applies in the southern hemisphere.) The wind shift can be helpful in determining when you have passed the front; conversely, if you are dead reckoning without any means of knowing that the expected wind has shifted, other signs of frontal passage, such as a change in temperature or type of cloudiness, indicate that the wind change has occurred and that it is time to make a heading correction.

In the later stages of the life of a low pressure system, the cold front may overrun the warm front near the low center, producing an occluded front with complex weather conditions. On weather maps, a cold front is represented by a line with pointed black excrescences, like thorns; a warm front replaces the thorns with misleadingly benign-looking semicircles; an occluded front uses the two symbols in alternation. Although symbology (the lines representing fronts) and terminology (expressions like "frontal surface") suggest that fronts are more or less distinct boundaries, they are not; they are broad zones of air mixing between the neighboring air masses. The breadth of the frontal zone will partly determine the extent and violence of the weather associated with the front, and the sharpness of the "kink" in the isobars which coincide with it. The wind blows *through* the front. A weather balloon released behind a cold front may materialize, some time later, in the warm air mass ahead of it, having passed through the front. Because of the penetration of the front by wind, there is continual mixing between the adjoining air masses; they retain their identity for long periods only because the low center continues to entrain warm and cold air from different geographical areas into the weather system, replenishing the supply of moisture and temperature difference which is constantly being dissipated by mixing.

Besides the local, fast-moving, familiar fronts produced by low pressure systems, there are large, semipermanent fronts in areas where tropical-, polar-, and temperate-zone air masses meet. The poles, for example, produce a mass of cold air at high pressure which presses outward along the "polar front," while in the tropics there is permanent atmospheric instability and much thunderstorm activity associated with the so-called "intertropical convergence zone" or "intertropical front." On the other hand, warm and cold fronts and their associated weather patterns don't occur in the tropics.

Satellite pictures provide a highly accessible picture of one aspect of the weather that interests pilots greatly: cloudiness. When fronts become occluded, or stop moving, forecasting is difficult; but the satellite picture immediately reveals what you are curious about on the simplest level: there are thick clouds here, thin clouds there, and no clouds somewhere else. This may not be vitally important information, but when all else is uncertain, it is at least comforting to know where the clear areas are.

Satellite pictures are available, in the United States, at U.S. Weather Service offices, at some Flight Service Stations, and at some major television stations. They arrive from geostationary satellites 22,000 miles above the earth, and take several different forms. For long flights, the first picture we are interested in is the 2-mile-resolution (wide angle) one in visible light. At the left top of the picture is a time and date. Above it is a gray scale which runs continuously in steps from black to white; this code, plus the naturalistic appearance of the photograph, tells us that this is a visible light photograph. It provides a general picture of areas of cloudiness and often of circulation, but little information on cloud heights, except what we might infer from the appearance of different areas of cloud.

The next type of picture to examine is the unenhanced infrared, which has a hazier appearance and a second band of gray scale below the line of alphanumeric code. The code contains the group "2ZA." In this picture, the temperature of the clouds corresponds to their shading, white clouds being colder than gray ones, and therefore at a higher altitude. An experienced analyst may be able to make a detailed interpretation of the infrared chart, but I usually just figure that the white areas are clouds above 10,000 feet and the gray areas are below that.

The last version of the picture is the computer-enhanced infrared coded "2EC": at first it looks as though the photograph has cancer, but what you are seeing is a kind of contour mapping which is achieved by using a gray scale which returns to black at two temperature intervals. The temperature contours are typically around −15 degrees and −50 degrees Celsius. The gray scale below the alphanumeric codes reads in diminishing temperature, and therefore increasing altitude, from left to right, with the zero point at the hash mark just left of the group of five numbers following the code "2EC." Each hash mark represents an interval of 10 degrees Celsius.

Visible light satellite picture of a classic Pacific Coast cyclone.

By themselves these contour lines are of no use, since they represent a temperature, not a height. But by consulting the winds-aloft forecast for the area where your track crosses a contoured area, you can find out at what altitude the air temperature is -15 or -50, and so determine the height of the clouds in the picture. In the standard atmosphere, a temperature of -15 degrees Celsius corresponds to an altitude of 15,000 feet.

Using the unenhanced infrared chart (code 2ZA), you again locate the zero point (the freezing level) on the lower gray scale, and compare the tone of the clouds in the picture with it. The unenhanced picture makes the location of areas of high buildups quite obvious, though it does not at first glance appear to be as crisp and precise as the visible light picture.

Given the underlying characteristics of low pressure systems, it is apparent that flights across a system toward the east, with the low on the left, will provide favorable winds but will lead toward a large expanse of cloudy, warm front weather. Flights toward the west will, on the other hand, offer a choice between keeping the wind on the tail but flying through an extensive area of bad weather to the north and east of the low (with lower temperatures and possibly worse icing conditions), or flying into the wind but to the south, toward the cold front. Expectations are modified by terrain conditions; a front moving across a body of water will pick up moisture and produce low visibilities and fog; motion across a mountain range will wring much of the moisture out of the air and usually will involve heavy precipitation on the windward sides of the mountains. If you consider that half the air in the atmosphere, by weight, is located below 18,000 feet, it is apparent that a mountain range like the Sierra Nevada or the Rockies, with heights of 10,000 to 14,000 feet, can make important modifications in weather characteristics.

Some areas have much more bad weather than others. The northwestern United States, for example, especially the area between the Great Lakes and the eastern mountain ranges, gets particularly severe winter weather as cold air flows down across the Canadian plains, picks up moisture from the lakes, and then dumps it on the windward slopes of the mountains. In planning a flight across that area in winter, or in fact across the United States generally, one almost always swings southward, crossing Ohio and West Virginia in preference to Pennsylvania and New York, or New Mexico in preference to Colorado.

On the west coast, the only area that regularly gets bad flying weather is the Pacific Northwest, which lies on the Pacific storm track and at the foot of the windward side of a significant range of mountains. It is rare for an Alaska-bound flight to follow the coast; most flights go up or down the inland valley, which is far drier. In conditions in which the coastal range is covered up solidly with layers of ice-filled clouds, the inland valley may well be broken or scattered. Farther to the south, only a few storms penetrate. This is a disadvantage for west coast pilots, at least from one point of view; when they go out to do battle with serious IFR weather, they are usually ill prepared. In the southern half of California, it is difficult to maintain instrument currency the year around without taking some pretty long trips.

The good weather of the southwest is dominated by a more or less permanent high located in the Pacific Ocean off the coast of Baja California. It is this high which prevents low pressure systems, which descend out of the Aleutian area and cross the coast between San Francisco and the Canadian border, from swinging far enough to the south to bring much precipitation to southern California. During the winter the Pacific high weakens and the Aleutian low grows stronger and moves southward, bringing a series of cold fronts into southern California and strong westerly winds to the route between San Francisco and Hawaii, which during the summer usually has a mixture of mild westerlies, easterlies, and northerlies. During winter it is often necessary to wait days for acceptable headwinds before flying from California to Hawaii, whereas during summer one day is generally as good as another.

A similar southward shift of low pressure occurs in the Atlantic during winter. Circulation there is dominated, in general, by a low situated around the southern tip of Greenland. Since the normal direct routes across the North Atlantic pass to the south of that low, there is often a tailwind on Europe-bound flights and a headwind on the return. Sometimes the low moves southward, giving tailwinds for the return and making a route via the Azores preferable, for reasons of wind or weather, for getting from North America to Europe. A typical situation on a westbound flight might find the low fairly far to the north, with a large area of cloudiness and precipitation extending from Ireland to west of Iceland, with relatively warm air, rain, fog, and little or no icing; and west of there, a post-cold-front type of weather, with an extensive cloud layer with tops at 6,000 or below, gradually becoming broken and scattered toward Canada, and fierce westerly winds—forty or fifty knots—on the nose off the tip of Greenland, becoming weaker and more northerly as you approach Goose. You would find icing in the cloud deck behind the cold front, and you would be obliged to choose between flying low enough to stay below the freezing level and avoid the worse winds, or high enough to stay out of the clouds altogether, despite the stronger wind. Over wide expanses of ocean, the wind gradient with altitude is not so steep as over land, and the benefits of seeking a lower altitude in a headwind are not necessarily so great.

Because of the complex influences of land and water on the Atlantic weather, crossings via Greenland and Iceland during the winter can be difficult, with a lot of unforecast bad weather; the oceans supply so much moisture to air masses that the usual rules of clearing behind fronts may not hold true. For this reason, if for no other, it's a good idea to have sufficient range capability to make the winter crossing by a more southern route. During the summer and early fall, however, the northern route is reliable enough, and it can sometimes be crisp and clear all the way between the storm systems in winter too.

Still, it would be rare to cross the North Atlantic, for example, and not see a cloud, for the simple reason that it's rare, anywhere on earth, for there to be no clouds at all over such a long distance. Long flights are practically certain to encounter weather, simply because they are long. For that matter, they are exposed to unplanned changes in en route conditions. To some extent, you simply take your chances. I recently took off from Los Angeles on a beautiful day to fly to Santa Fe, New Mexico. The weather briefer offered clear skies and tailwinds for the 620-nm trip. The tailwinds were there, and I arrived at Santa Fe in time to make an approach to minimums in a snowstorm. The weather was a 400-foot ceiling and one mile in snow and fog. This rather startling change took place in the course of a three-and-a-half-hour flight. It is an extreme case, but it illustrates how wrong weather forecasts can be, and how important it is to stay abreast of terminal weather.

It also illustrates another point. The snowfall in Santa Fe did not come out of the ground; it was spawned by a larger area of bad weather around Denver, which a high in New Mexico was supposed to be holding in check. The high leaked. A quick briefing, covering the surface observations at ground stations along my route, revealed nothing about the system to the north. Not that the sudden prolapse was foreseeable, but it wouldn't have hurt to know that there was a threat of bad weather to the north, and that it might break through. The regional weather picture really can matter; if you can't get a look at a weather map first hand, you should still get clear in your mind the locations of highs, lows, fronts, and areas of precipitation to both sides of your route, before listening to the litany of terminal forecasts and sky conditions.

Although heavily traveled international routes like the North Atlantic and the central Pacific have good weather coverage, and forecasters and briefers of a high order, detailed weather information is not necessarily available in countries less thoroughly developed, or less densely populated, than the United States. Even when information is available, you may not be able to get the benefit of a local forecaster's experience, because you can't speak his language. You have to settle for a few terminal surface observations and hope for the best in between. You can get some surprises. There may be a range of mountains between here and there, for instance, that gets completely covered by cunims every afternoon. All the local pilots know it, but you don't, and nobody may tell you.

There is a final twist to foreign weather: the form of the reports. Other countries use a different format from ours and a different symbolic language. It's quite a stew. Some of the letter codes are clear enough (CU means cumulus; HZ, haze; RA, rain; FG, fog), but some are less obvious (GR is hail—*grêle* in French; BR is mist—*brume* in French; FU is smoke—*fumée* in French; BC is "patchy"—I don't know why). Visibilities are given in meters (or kilometers), wind speeds are in knots; but ceiling heights are in feet, because altimeters read in feet all over the world. (Nevertheless, the pressure-surface heights on upper level weather charts are in meters, not feet—right?)

Conditions restricting visibility are broken down by a number and letter code, and when there is a ceiling the cloud type is also coded. The cloud types are simple enough: ST is stratus; SC, stratocumulus; NS, nimbostratus; and so on. There are only nine cloud types, and if you know your cloud types you won't have any trouble interpreting the initials. Visibility details are harder to interpret. The numbers begin with the range from 10–49, which indicates the degree of obscuration and that the obscuration is due to small particles—vapor, dust, haze, or fog. It seems remarkable that anyone can discriminate forty stages of obscuration, but there it is. Light rains are coded 50–59; moderate rain from 60–69, and heavy rain, hail, or snow is 90–99; 70–79 means snow, and 80–89 means showery, intermittent precipitation of any type. Two-letter codes following these numbers give further clarification.

The report, which is called a TAF (Terminal Aerodrome Forecast), begins with four numbers which indicate the time of its validity (2121—the 24-hour period from 2100 Zulu one day to the next); successive paragraphs of the TAF begin with codes like TEMPO, GRADU, INTER, or PROB, meaning that the conditions in the paragraph will occur temporarily during the period of the forecast; that they represent a gradual trend; that they will be intermittent; or that the forecaster sees so-and-so much probability (expressed in a percentage) of the condition occurring.

The following is an example of a cheerful forecast for a summer day in Shemya, Alaska. It begins with the station identifier, PASY.

```
PASY  2121 24020/30 0800 50DZ 45FG 9//002 QNH2944INS CIG002
      GRADU 0607 24022/32 0600 61RA 45FG 9//002 QNH2948INS
                                          CIG 002
      GRADU 1415 26020/30 1600 60RA 44FG 6ST006 8SC008
      QNH 2944 INS CIG003 COR 2141
```

The notation 9//002 means indefinite ceiling 200 feet; 6ST006 means "six octas stratus, ceiling 600," or, as we would say in the United States, 600 broken. The octa is a measure of sky area; it means eighth, and so six octas is three-quarters obscuration. QNH means "barometric pressure"; "INS" is inserted after it because this is an international report on a U.S. station which reports its QNH in inches, not millibars.

This midsummer report explains why no one goes to Shemya who can go to prison instead.

To confuse matters, the Canadians have their own weather reporting language, generally similar to that used in the United States.

It bothers me sometimes that weather reports omit to give the hours of sunrise and sunset. As far as a pilot is concerned, that information can be just as useful, on long flights, as the visibility and the ceiling. If Shemya is going to go from a 2000-foot RVR with indefinite low ceiling to 600 broken, 800 overcast, and a mile at 1400 Z, which is two o'clock in the morning, I would be curious to know when the sky will start to grow light. In Alaska at that time of year it could be at two in the morning, for all I know. Terminating an approach to minimums at an isolated runway in total darkness is something like landing on an aircraft carrier at night: you aim toward an illuminated hoop in black space without much conviction of its real size, position, or solidity. When you're worn out, a little light can be a great help; in fact, it seems to me to be an essential part of the weather picture, as far as a pilot is concerned, and I'm surprised that they don't give recognition to it in the sequence reports. Sunset and sunrise tables can be obtained, however; they are included in Jeppesen trip kits.

6

EFFICIENT FLYING

People use the word ''efficient'' in a hundred different ways, sometimes interchangeably with ''effective'' (as in ''This is a very efficient cheese-grater''), and often to mean nothing more than ''good.'' But efficiency is a scientific concept of precise application, and when I speak of efficient flying I mean to use the word in its scientific sense. It means getting the most work done in exchange for the least fuel or cost. For our purposes, it boils down to the most distance for the least fuel. This is a definition of efficiency that makes distance (rather than speed) the essential measure of performance; obviously it is the one which is of the greatest concern to the long-distance pilot, and it is of growing concern to all.

Airplane efficiency may be measured in miles per gallon—nautical miles per gallon, or nmpg, henceforth, because IFR flight plans will be filed, and chart distances read, in nautical miles. Alternatively, it can be expressed in terms of absolute range—the distance an airplane can go by burning all the fuel in its tanks. Absolute range is the product of miles per gallon and gallons burned, and so the two measures can be used interchangeably to express efficiency. I will use absolute range as a yardstick, however, because it can be calculated by means of a rather simple equation called the Breguet range equation, and by examining this equation in detail we can gain an understanding of the variables affecting efficiency.

The Breguet range equation yields an absolute range based on a small set of variables. It ignores fuel used to taxi, take off, and climb, and considers only a hypothetical airplane launched instantaneously in flight with a certain fuel load, and allowed to consume all that fuel while always flying at its most efficient speed. The concept of most efficient speed is of central importance, and I will return to it later.

The equation is as follows:

$$\text{Range} = 751 \times \frac{\eta \times L}{c \times D} \times \text{Log}_{10}\frac{W_1}{W_0}$$

where η = propeller efficiency
c = specific fuel consumption, pounds per horsepower per hour
L/D = Lift/Drag ratio (maximum)
W_1 = Takeoff weight
W_0 = Takeoff weight less fuel burned

The number 751 at the start of the equation is a dimensional constant. The propulsive efficiency, η (pronounced "AYta"), is a measure of the portion of engine power that actually ends up doing useful work. The maximum value is usually around .85 or .86; fixed-pitch propellers may have considerably lower efficiencies than this—.75 or less—but constant-speed props usually do not drop below .80 in cruising flight. If one is curious about prop efficiency, one can call a prop manufacturer, give the design number (or airplane model), name an rpm, cruising speed, altitude, and power setting, and get, in exchange (with a little coaxing) a reasonably accurate, but of course theoretical, figure. An error in estimating prop efficiency will not change the final result by more than a small percent, however; so in addition to being hard to determine, the prop efficiency is of comparatively little concern.

Specific fuel consumption (sfc) is the number of pounds of fuel used by the engine per hour to produce one horsepower. The most efficient aircraft piston engines have fuel specifics of around .40–.45 lb/hp–hr with a lean mixture. For example, an engine rated at 200 hp uses 10 gph at a lean mixture at 75 percent power. Ten gallons of fuel weigh 60 pounds, and 75 percent of 200 is 150; 60/150 equals .40, the specific fuel consumption.

Enriching the mixture increases the fuel specific; a specific of .6 pounds/hp per hour might be used in climb, when excess fuel is needed to keep the engine cool. Old-fashioned engines, like the Continental 0–200, have poor fuel specifics—not much better than .50 to .60 at best, even with a lean mixture; so do very powerful turbocharged engines, and gas turbines. In general, injected engines have better fuel specifics than carbureted ones, and turbocharged engines have worse specifics than normally aspirated ones.

Note that if the difference in specific fuel consumption between climb and cruise is nearly 50 percent, it is obviously important to attend to proper leaning of the mixture for efficient cruising. In fact, this is one of the most significant elements in the range equation.

The ratio L/D, the lift-drag ratio, varies with the angle of attack. It is low at high angles of attack, near the landing speed, and at low angles of attack, around the cruising speed; it reaches its maximum value at an intermediate speed which is typically somewhere near the speed for best rate of climb and which is called the "best L over D speed," or $V_{L/D}$. This is an indicated airspeed, not a true airspeed, and it varies with weight, since what is really affecting L/D is angle of attack, not speed. Speed is simply a convenient handle for grasping angle of attack, in the absence of an angle of attack indicator.

$V_{L/D}$ is the indicated airspeed at which an airplane goes the most miles on a gallon of fuel. It is not to be confused with V_E, the maximum endurance speed (also called V_{MP}, the speed for minimum power), at which the plane maintains altitude with the least fuel flow, and can stay aloft for the longest time.

Values of L/D for general aviation airplanes range from a low of perhaps 8 to a high of 15. Some homebuilts, such as the Rutan *Long-EZ,* have L/D ratios as high as 20. Values for sailplanes are much higher—up to 50. In general, a clean airplane with relatively a high aspect ratio wing, like a Mooney 201, will have a high L/D ratio, whereas a draggy airplane with a stubby wing (like a Tri-Pacer, for instance) will have a low one.

The final element in the range equation is the expression $\log_{10}\frac{W_1}{W_0}$. This is the logarithm, to the base of 10, of the ratio of the starting weight and the finishing weight for the trip. For ordinary light aircraft, the fuel supply accounts for perhaps 10 percent of the gross weight; long-range tanks may bring it up to 20 or 30 percent. The weight ratio is thus usually between 1.1 and 1.5, and its logarithm between .04 and .17.

An efficient single-engine airplane, like a Mooney or a Bonanza, might have a L/D_{max} of from 12 to 15 or better. Assuming an L/D of 15, .85 prop efficiency, and .42 specific fuel consumption, the product of all the parts of the Breguet formula except the log of the weight ratio is about 22,800. I'll call this the "efficiency index." Assuming a 2,000-pound empty weight and 500-pound cabin load (a couple of people and 200 pounds of baggage and equipment), therefore a W_1 of 2,500, one can easily work the Breguet formula backwards to find out the gross weight (and therefore the amount of fuel) required to achieve certain ranges. A range of 1,000 nm, for instance, requires a weight ratio logarithm of .0438, whence a weight ratio of 1.106, a gross weight of 2,765, and a fuel supply of about 44 gallons (assuming, remember, no climb, no reserves, and perfect maximum-efficiency cruising). A range of 2,000 nm requires 93 gallons; 3,000, 147 gallons. The reason the fuel required increases faster than the distance traveled is that the more fuel you take aboard in the first place, the more you weigh, and so the greater your initial fuel flow.

A lesser value of L/D or prop efficiency, or a higher specific fuel consumption, would reduce the efficiency index and consequently increase the fuel requirement. For example, suppose that the L/D were 12, the prop efficiency .80, and the fuel specific .45, as might be the case with a carbureted, fixed-gear single with a fixed-pitch prop. The efficiency index would drop to about 16,000. Suppose, however, that the airplane's zero-fuel weight is lower, perhaps 2,200 pounds. In this case, a 1,000-mile Breguet range requires 57 gallons; 2,000, 122 gallons; and 3,000, 198 gallons.

Finally, take the case, still hypothetical, of a proposed round-the-world nonstop flight by a light airplane. Suppose that the airplane is Jim Bede's *LOVE* (Low Orbit Very Efficiently), a Schweitzer 2–32 sailplane fitted with wing tanks and a 210-hp Continental engine, and that it has an L/D of 25, an η of .85, an sfc of .48 (because much of the trip must be spent at an extremely low power setting, at which the engine is not capable of its best efficiency of .40–.42), and a zero-fuel weight of 2,000 pounds. (These figures are picked at random; I don't know the actual values for the Bede project.) The efficiency index is 33,247. Assume that the distance to be flown is 18,000 nm; the fuel requirement will be 908 gallons. Evidently Bede made some very optimistic assumptions, since I doubt that 908 gallons could fit into a 2–32's wing, or that the engine could get the 7,400-pound airplane into the air. If the L/D were 30, the fuel specific .40, and the distance 15,000 miles, the fuel requirement would drop to 388 gallons—a more manageable value.

It is very interesting to note that the Breguet range equation hinges upon a weight *ratio,* not an absolute weight. In effect, it expresses the fuel requirement in terms of the takeoff weight. A one-pound airplane requires half a pound of fuel to go the same distance as a two-thousand-pound airplane can go with a thousand pounds of fuel, or a million-pound airplane with half a million pounds of fuel, other things being equal.

You can see from this series of examples that the range of an airplane is influenced by relatively few factors. Of these, even fewer are within the control of the pilot; namely, his speed, mixture, and fuel supply. A fourth might be mentioned, since some pilots place more reliance upon it than upon the others: that is, wind. A flight of long duration can benefit considerably from a tailwind, a 20-knot tailwind increasing the distance flown in ten hours, for instance, by 200 nm. For a 200-knot airplane this is not so significant as for a 100-knot airplane, to which it represents two hours' flying.

The next question would necessarily be how strongly different factors affect fuel requirements. In the case of the Bede *LOVE,* for example, reducing the fuel specific by 17 percent and increasing the L/D by 20 percent, while shortening the trip, which is tantamount to assuming a tailwind, by 17 percent, reduced the fuel requirement by 57 percent. This appears encouraging, but it cuts both ways; making incorrectly optimistic assumptions about efficiency and wind might cause one to fall 57 percent short of the required fuel supply.

Taking a 2,500-pound zero-fuel weight, you can see how it is influenced by the efficiency index. On the graph, (next page) efficiency indexes from 10,000, which represents the least efficient airplane anyone would ever dare take on a long trip, to 30,000, which represents the best, are plotted against the fuel requirements for a 2,000-mile trip (representing the Atlantic crossing, for instance, from Gander to Shannon). Note that as the efficiency drops, the fuel requirement increases drastically.

A value of about 22,000, remember, would represent a clean, retractable fuel-injected airplane flown at its most efficient speed. Remember too that the equation actually deals with a weight ratio, in which the fuel required is expressed in terms of the weight of the airplane. Thus, a 5,000-pound airplane requires twice the fuel that a 2,500-pound one requires, all other things being equal.

Now, the three elements affecting the efficiency number are η, c, and L/D. Specific fuel consumption may vary, between a best-power mixture and a best-economy mixture, by 20 percent. A similar change in the efficiency number, reducing it, say, from 22,000 to 18,000, increases the fuel requirement by about a quarter in our example.

The best L/D ratio of the airplane cannot be influenced by the pilot; but the cruising L/D can, by his selection of a cruising speed. Obviously there is no reason to wish to fly at a speed lower than $V_{L/D}$; but in the interest of comfort one might well wish to fly faster.

Flying faster than $V_{L/D}$ (ignoring for the time being the effects of wind) causes a diminution in efficiency, but the rate of the diminution is relatively slow. Flying 10 percent above the $V_{L/D}$ reduces the cruising L/D by less than

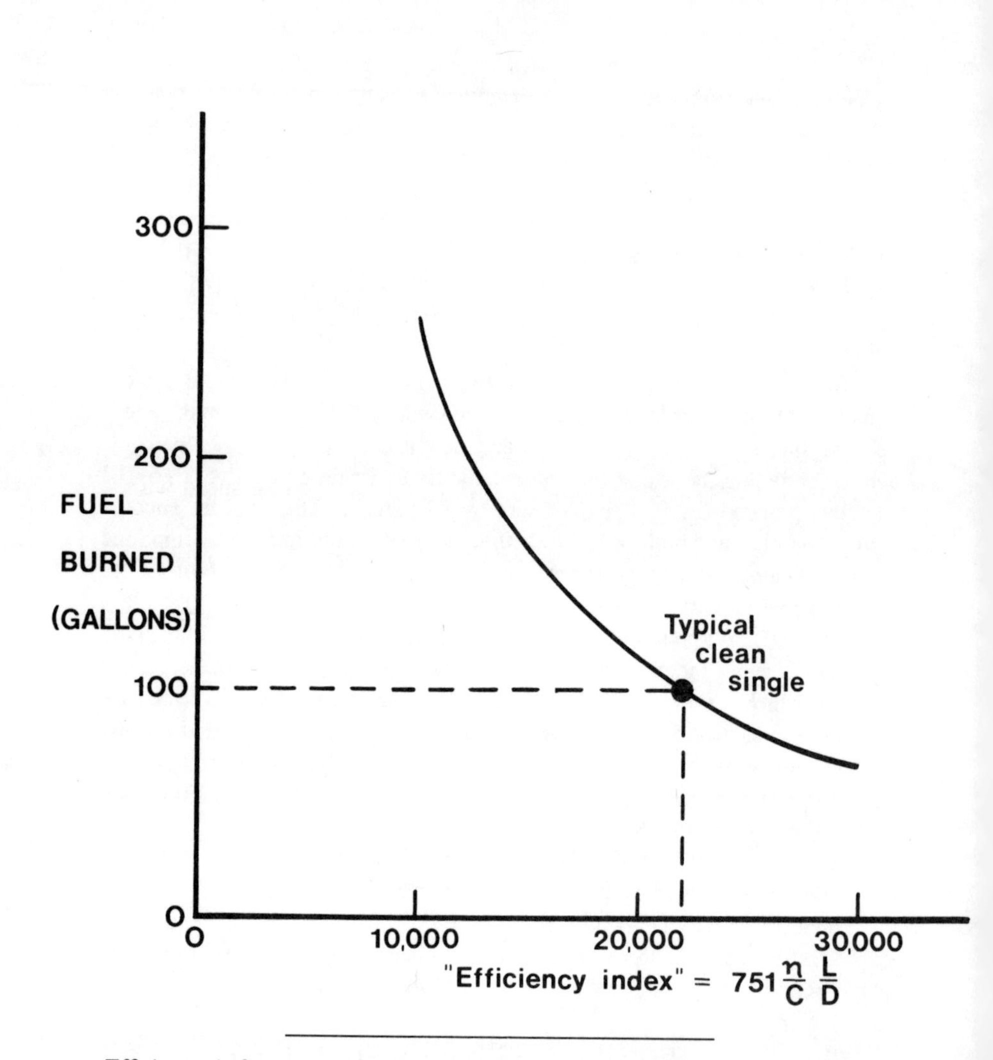

Efficiency index versus fuel required for a 2,000-mile trip, for an airplane with a 2,500-pound zero-fuel weight.

3 percent for a typical clean light airplane; 20 percent above reduces it by 7 to 8 percent; 30 percent by around 15 percent; 40 percent by 20; 50 by 27; and so on. Generally, the longer the wingspan and the higher the maximum L/D, the more deleterious the effect of flying off the most efficient point; but the differences between airplanes of normal proportions and aerodynamic cleanness are very slight.

As we saw before, however, the Breguet fuel requirement increases somewhat more rapidly than the efficiency index diminishes; and so flying 20 percent above $V_{L/D}$, while it may decrease the value of L/D (and therefore the range constant) by 8 percent, might increase the fuel required for a given distance by 10 percent. The effect is progressively greater as the efficiency index shrinks, as the graph above showed.

Since the most efficient speed is the one at which distance covered is greatest compared with fuel flow, it is apparent that the best speed will be affected by wind, because wind affects the distance covered. A tailwind, by increasing distance covered at no cost in fuel, pushes the most efficient speed downward; a headwind pushes it upward. Without going into the mathematics of the matter, it is enough to remember the approximate rule that around $V_{L/D}$, the indicated airspeed should be increased by one quarter of the headwind component, and diminished by one quarter of the tailwind component to attain the peak values. This correction applies only at the most efficient speed. If you are cruising at a much higher indicated speed, the increase required by the headwind will be much smaller.

Normally you would not set off on a trip which would be outside the normal range of your airplane without a tailwind; and so the technique of reducing speed for best efficiency with a tailwind is of academic interest, though it could be useful in a distance record attempt. Minimizing the effect of headwinds may be of more practical importance. It is important to distinguish, however, between the particular L/D you are getting and the best one you could get. Suppose you are cruising along at 70 percent of power, and you discover that an unforecast headwind is slowing you down and may prevent you from reaching your destination. The remedy is not to speed up, because you are already probably flying well above your most efficient speed. The remedy is to slow down. The effect of the headwind is merely to raise the *most efficient* speed by one quarter of the headwind component—not all speeds. The entire curve is shifted upward by a headwind; the best efficiency is in any case worse than in still air. In an extreme case—a seventy- or eighty-knot head wind, for instance—the most efficient speed may become the maximum speed, and any lower speed would be less efficient. At the same time, the significance of speed changes is diminished by a headwind. The greater the headwind, the less it matters how much above $V_{L/D}$ you fly.

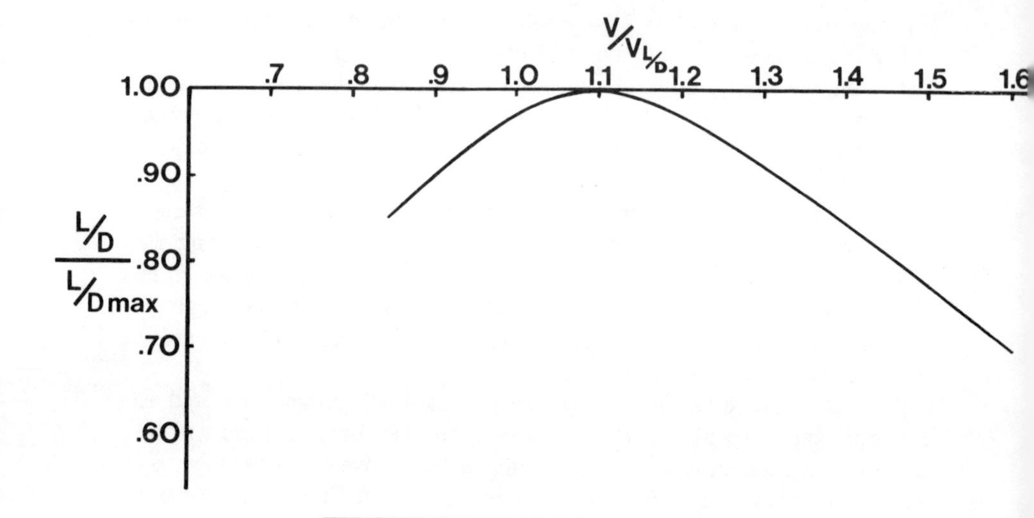

Variation of L/D with speed.

All this may be made clearer by a graph which represents the relationship of fuel flow and groundspeed. Since we are interested in the ratio of the two variables, the feature that concerns us is the slope of a line drawn from the origin (that is, the lower left hand corner) to intersect the curve at a given speed or fuel-flow point. This curve, while its precise shape differs from airplane to airplane, is representative of the characteristics of all airplanes.

To display the effect of a headwind, we have only to move the origin, leaving everything else the same. For a forty-knot headwind, for instance, we move the origin forty knots to the right. The speed for best range, which is the one at which a line from the origin is tangent to the curve, is now higher, and the slope of the line is also greater. In other words, the nmpg figure is worse, and its peak occurs at a higher speed.

The farther to the right the origin moves, the closer the point of tangency moves to the maximum speed. Furthermore, the point of tangency is moving into a portion of the curve which is relatively flat; that is, there is little difference in slope (i.e. in nmpg) between a line which is tangent to the curve and one which intersects the curve at some distance to either side of the point of tangency. In other words, with a large headwind component, efficiency is insensitive to speed.

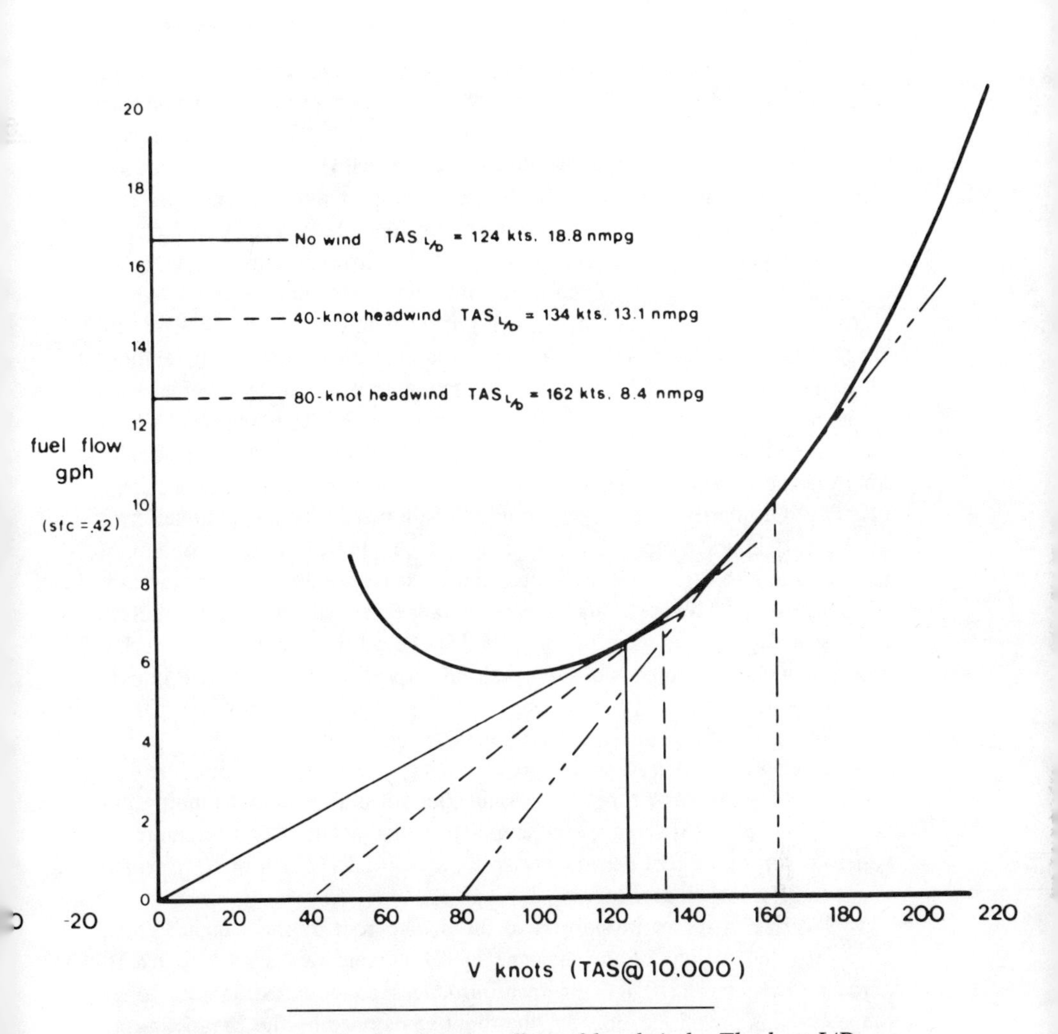

Fuel flow versus speed, showing the effect of headwinds. The best L/D, or maximum range, speed is where a line from the origin grazes the curve. A headwind shifts the origin to the right, raising the max-range speed. For moderate headwinds, the speed increase is about a quarter of the headwind component.

So far I have been discussing only theoretical range figures. In planning actual flights and evaluating the supplementary fuel requirements which they might involve, you must take into account all sorts of additional margins.

To begin with, airplanes must climb to a reasonable cruising altitude; and they must have some reserve aboard on arriving at their destination. Two hours at normal cruise speed is the international standard for overwater fuel reserves; obviously, if intermediate airports were available the reserve could be smaller. The range involved in the Breguet range equation is called the "gross still air range," and it is much greater than the safe range. The safe range must include, besides allowance for climb and reserve at landing, allowances for headwinds, possible navigational errors, and problems, such as the loss of an engine, that might occur during flight. The headwind allowance is arbitrary, since, obviously, the owner who equips his airplane for long-distance flights does not know what the headwind will be on a particular day. Ordinarily, one might include the allowance for navigational error in the two hours' reserve, which can also be applied to the problem of engine failure in flight (not a problem, in the sense we are now considering, for single-engine planes). Inadequacy of range to deal with engine failure may be acceptable in view of the fact that the portion of the flight during which it would be impossible to reach an airport after loss of one engine may be quite small, and the chance of an engine failing during that particular stretch is even smaller than that of its failing at all.

As a practical matter, I would prefer a reserve at landing of three or four hours, rather than two, because I would like not only to avoid running out of fuel, but also to avoid even having to think about it. Furthermore, a generous surplus of fuel permits one to fly at a speed well above $V_{L/D}$, which is apt to be frustratingly low.

$V_{L/D}$ varies in direct proportion to the square root of the weight. Thus, increasing the weight of the airplane by 21 percent raises its $V_{L/D}$ by 10 percent ($\sqrt{1.21} = 1.10$). This fact introduces a complication into long-range cruising technique. Ideally, the flight envisioned by the Breguet range equation takes place at a constant L/D (that is, a constant angle of attack), and therefore at a speed which continually diminishes as fuel is burned and the weight of the airplane is reduced. While adhering to such a regime would impose no undue intellectual burden upon the pilot, who would merely reduce his speed to some precomputed value each hour at the same time as he changes his heading to match a great circle route, it would tax his patience, since at the end of the flight he would be crawling along at a relative snail's pace while with all his heart he wished to hasten at last to his destination. After ten hours in the air, one wants to go faster, not slower.

As we have seen, the effect of deviating by 10 or even 20 percent from the best speed is relatively small. So it is customary in long-range flying to hit upon a single speed which compromises between the takeoff and landing weights; this speed may simply be the speed for the initial weight, on the premise that the deterioration of L/D ratio later in the trip will be applied to a smaller weight, and will therefore account for a smaller amount of excess fuel than if it occurred at the beginning of the flight, when the airplane's weight is greatest.

My own inclination has been to calculate all ideal speed and range information while planning a trip, write it all down, and then proceed at the highest speed that seemed practical (usually about 65 to 70 percent of power) while monitoring fuel consumption and apparent surface winds to the extent possible. On the one occasion when it seemed possible that I had made a navigational error and might be threatened with running out of fuel, I knew that I could slow down to best range speed when necessary and make a great deal of distance on my remaining fuel. The problem was deciding when to do it. On this particular occasion—twelve hours out of Alaska and a couple short of Japan, longitude uncertain—I never did reduce speed. But I took comfort afterward from the fact that my reserve at landing—23 gallons—would have been much more useful at the end of the flight than at the beginning, since it would have taken me some 570 nm at a light weight, and only 380 at a heavy one.

If you are going to cruise at a speed above $V_{L/D}$, then the choice of an altitude comes into play. You may have noticed with surprise that the Breguet range equation makes no mention of altitude, implying that the range of an airplane is not affected by altitude. This seems to contradict the common observation that a better true airspeed can be had at the same power setting by going to a higher altitude. If the speed is better at altitude when the fuel flow is unchanged, would not the range also be greater at the higher altitude?

The catch is that the Breguet range equation assumes cruising at $V_{L/D}$, while common experience is of airspeeds considerably higher than this. At $V_{L/D}$, true airspeed is directly proportional to fuel flow, because $V_{L/D}$ represents a unique angle of attack. Fuel flow increases with increasing altitude while indicated airspeed (IAS) slowly decreases and the increase of true airspeed keeps precise pace with the increase in fuel flow.

If, however, you fly at a rather high true airspeed, and climb to reduce the indicated airspeed (and increase the angle of attack) while keeping true airspeed constant, you will be moving toward $V_{L/D}$, and consequently into an increasingly efficient mode of flight. Because of the increasing efficiency, you get more miles per gallon, at a given *true* airspeed, at high altitude than low. This is only true, however, when the indicated airspeed is greater than $V_{L/D}$; the closer the IAS is to the best range speed, the smaller the benefit of higher altitude becomes.

If it is a foregone conclusion that you are going to cruise at a speed above $V_{L/D}$, then you should climb to the highest practical altitude to do so. With a heavily loaded airplane, in fact, you may find that at eight or ten thousand feet your maximum cruising speed *is* $V_{L/D}$, and in that case you will have the comfort of knowing that you cannot do any better than you are doing.

A turbocharged powerplant makes it possible, generally, to achieve simultaneously maximum aerodynamic efficiency and high speed. You have only to climb to the altitude at which a high cruise speed and power setting correspond to an indicated airspeed equal or close to $V_{L/D}$, and you reap the twin benefits of speed and economy. There are drawbacks, however. One is that if it is necessary to use oxygen continuously, the economy of efficient flight is obliterated by the high cost of oxygen refills. Another is that an airplane's oxygen range is usually inferior to its fuel range, so that the one, not the other, limits the length of a trip. In this case, one might climb to oxygen altitude for only part of the trip, and make that part coincide with the best stretch of forecast winds.

Another disadvantage of turbocharged engines is that they are generally somewhat less mechanically efficient than normally aspirated ones. They have lower compression ratios, the turbocharger itself is relatively inefficient, cooling drag is increased, and a richer-than-ideal mixture may be required during the climb and even in cruise in order to stay within permissible temperature limits and cool the engine at high altitudes, where the air cools poorly because, though cold, it is not dense. Excessive fuel consumption is a common complaint with turbocharged engines, contrary to the impression created by advertisements and articles suggesting that they give something for nothing. Finally, the turbocharger is one thing more (or two, or four) to go wrong.

The advantage of the nonturbocharged engine, looking at it stricly from the standpoint of efficiency, is that by climbing to 10,000 or 12,000 feet the pilot compels himself to fly efficiently; no further effort is required. A turbocharged engine complicates the picture, obliging the pilot to discipline himself and not make use of all the power and speed available. On the other hand, the turbocharged engine permits some fancy footwork in selecting power settings. Some power is wasted in overcoming the friction of engine rotation; this power may be measured by setting up the engine on a test stand and rotating it, at various speeds typical of cruising flight, with an electric motor. The spark plugs are removed so that the engine does little work pumping air; the work is principally done in moving oil films about. Friction being proportional to the square of speed, the friction horsepower increases rapidly with rpm. In the Continental 10-360 engine, for which I happen to have figures available, the friction horsepower is 14 at 2,100 rpm and almost 28.5 at 2,800 rpm. These losses are taken into account in published engine performance data, so that rated power outputs and fuel specifics represent net values; and since fuel specifics are never published against rpm in such a form as to make the information useful for pilots, I cannot say with certainty that low rpm and correspondingly low friction horsepower make for better fuel consumption, although this has been a traditional prescription.

If one accepts that low rpm is desirable, then manifold pressure is the only means available to raise the power output, and with a normally aspirated engine, high manifold pressures are available only at low altitudes. With a turbocharger, on the other hand, high manifold pressure can be combined with low rpm at any altitude.

Experiments in *Melmoth* have shown a striking improvement in fuel efficiency—as much as 5 percent—in reducing rpm from 2,500 to 2,400, a slight further improvement in dropping to 2,300, and no further improvement down to 2,000 rpm. The biggest differences seem to occur at high power settings; at the very low power settings corresponding to $V_{L/D}$ at low or moderate altitudes, the effect of rpm appears to be slight.

I have it on good authority that my particular engine—a TS10-360—is most efficient at 2,400 to 2,500 rpm because at that speed it achieves the most even mixture distribution. At lower engine speeds one cylinder may get significantly less fuel than another, so that some go excessively lean while others are still on the rich side. When my engine was normally aspirated I always leaned it by reference to the number five cylinder, which experiment had shown to be the first to arrive at peak EGT as the mixture was leaned. Now that it is turbocharged I lean according to the TIT, or turbine inlet temperature. This represents a collective EGT for all six cylinders, and it has a rather flat peak because as some of the cylinders are going over to the lean side others are just coming up to peak. I lean to peak TIT and then continue leaning until I detect what I consider an unacceptable level of roughness; I then enrich slightly to eliminate the roughness, and run at that mixture setting.

Turbocharged engines have a TIT limitation—1,650 degrees Fahrenheit—which puts a constraint on fuel economy at high power settings. Above 65 or 70 percent of power, peak EGT may produce TITs above the redline value, obliging you to enrich the mixture and therefore to increase your specific fuel consumption. You can also lower the TIT by leaning beyond peak EGT to the lean side; most engines become rough before the TIT comes back down to acceptable levels, however.

Many pilots have heard of something called "the square rule," which requires that the number on the tachometer—the rpm in hundreds—not drop below the number on the manifold pressure gauge—the manifold pressure in inches of mercury. In other words, if you have a manifold pressure of 23 inches Hg, you should not let the rpm get below 2,300. In aid of this, one is always taught to reduce throttle first, rpm afterward. The purpose of this rule is to protect the engine from detonation.

Detonation is a condition in which the fuel-air charge in the cylinder, or some portion of it, explodes spontaneously before it is ignited by the spark plug or by the advancing flame front which the spark plug triggers. It is a potentially destructive condition; and although when it occurs in automobiles it produces an audible "knock" or "ping," it does not seem to be audible in airplane engines. Unfortunately, the conditions which correspond to most efficient engine operation are also those which are conducive to detonation: high manifold pressure, low rpm, and a lean mixture.

The so-called square rule represents an extremely conservative attempt to deal with the problem of detonation in all engines. In reality, different engines show different degrees of susceptibility to detonation. Compression ratio is an important influence; the higher the compression ratio, the greater the likelihood of detonation. At any rate, engine manufacturers publish a power chart for each type of engine which shows the recommended operating range. Almost invariably, the recommended operating range violates the square rule; for instance, most engines can be operated at 2,100 rpm and 24 inches Hg without risk. In fact, even these charts are quite conservative, taking into account the worst combination of high cylinder head temperatures, low fuel octane (within the required range, of course), and lean mixture, retarded spark, and so on. It should be a basic part of your familiarization with an airplane to consult the power chart and commit to memory its extreme limits.

Good engine instrumentation is helpful in allowing you to operate in the most efficient modes. Conservatism in leaning the mixture and running very rich during climb are unnecessary if you have cylinder head temperature information to assure you that you are not over-temping the engine. It is far better to have indications for all cylinders than for one only; the rule that there is one cylinder which always runs hottest is not true, and many engines cool differently during the climb than in cruise, and at one power setting than at another. The same can be said of exhaust gas temperature information; there is no single leanest cylinder, unless you always run at the same power setting, altitude, and rpm.

Ferry pilots make use of some unconventional leaning techniques to achieve maximum fuel economy. The best ratio of fuel flow to power output occurs at a mixture setting at or just to the lean side of peak EGT, while the peak power output, at a given throttle and rpm setting, occurs somewhat to the rich side of peak. One technique of leaning an engine with a fixed-pitch prop, as on a Cessna 172, would be to climb to cruising altitude, say 8,000 feet, and then to lean to reduce to cruising power, leaving the throttle wide

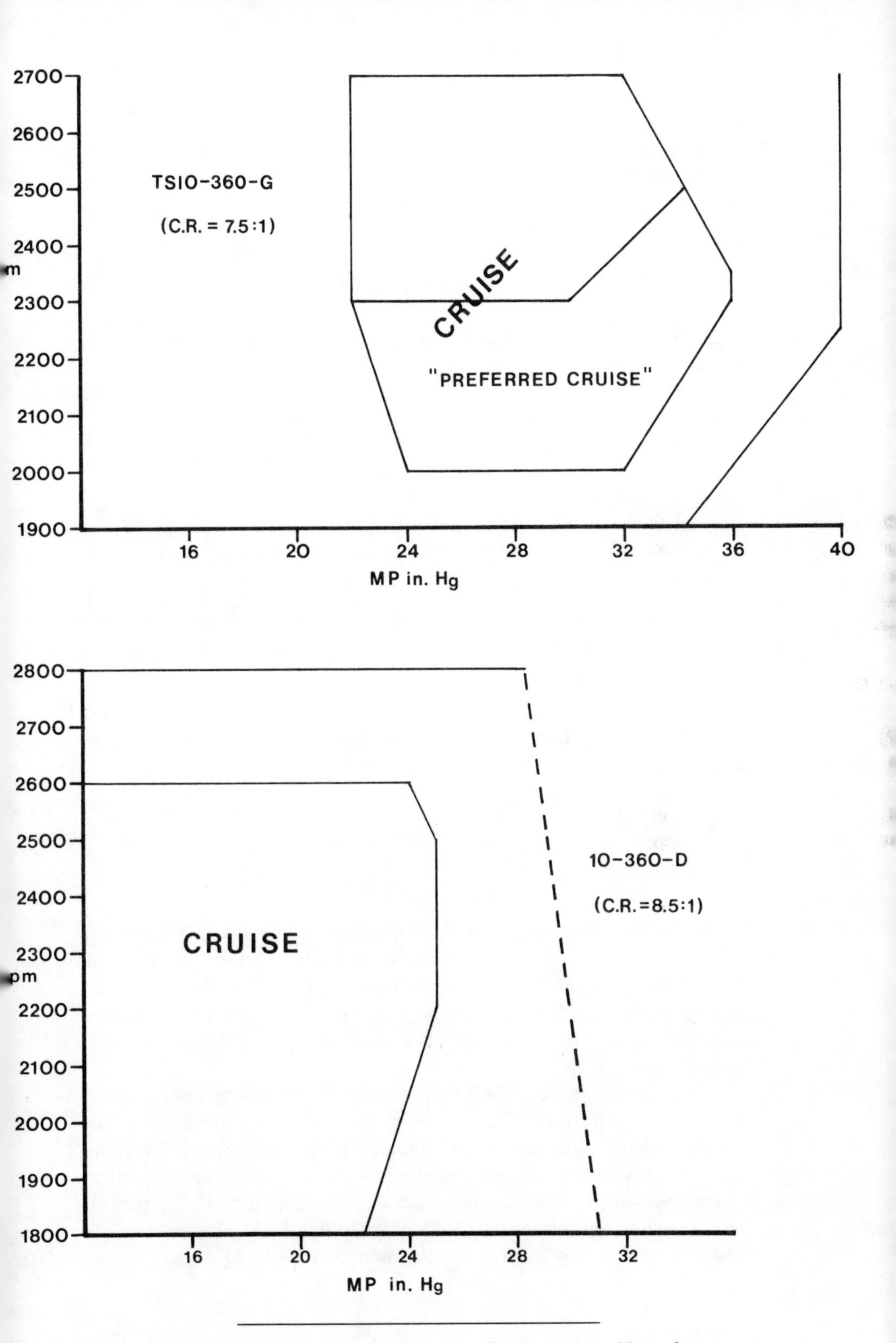

Power charts for two versions of the same basic engine. Note that over-square settings are permissible even in the high compression, normally aspirated version.

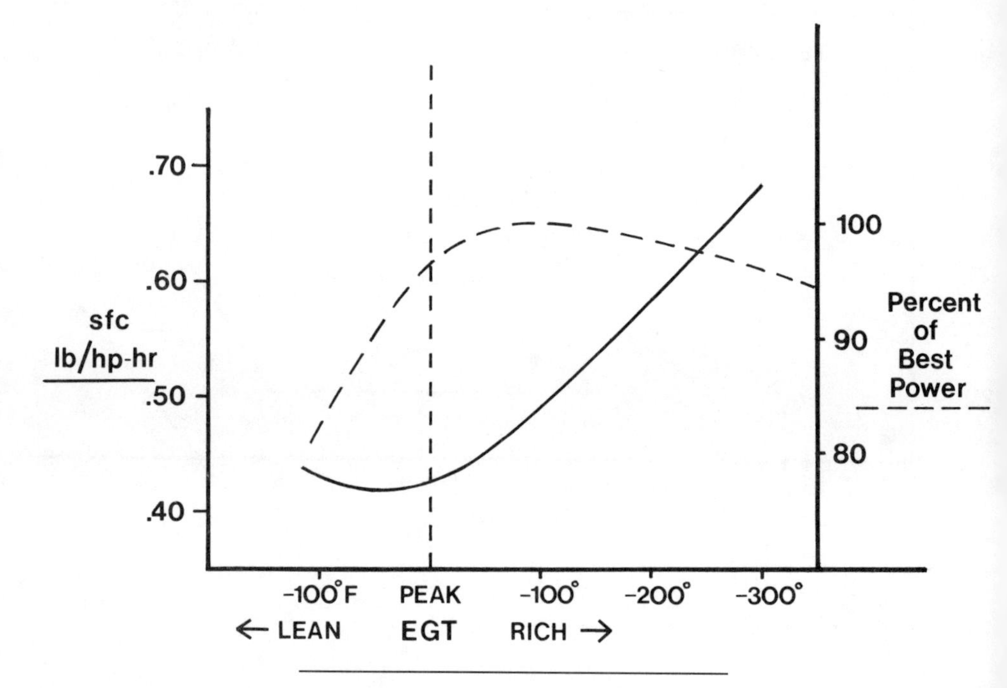

EGT versus specific fuel consumption. The most efficient or economical mixture is at or somewhat to the lean side of peak EGT.

open. After an initial rise in power (and therefore in rpm) the rpm would drop to the desired value, and the mixture would be at peak or on the lean side. While this technique does not guarantee any particular power output or economy, it implies leaning through a significant drop in rpm, and that means that the engine will be operating in a very lean, and therefore very efficient, mode.

With a constant speed propeller, it is necessary to lean by engine sound rather than by prop speed. The trick in this case is to lean until a noticeable roughness appears, and then to vary throttle and rpm settings to try to eliminate the roughness. Max Conrad used to say that he would lean his engines out until they were barely running, and that in addition to saving fuel he would thereby protect himself against plug fouling. I am not sure I understand why there is any point in leaning more than 50 to 75 degrees on the lean side of peak; but I have observed again and again that when descending from altitude without enriching the mixture I can bring about mixtures 100 degrees or more on the lean side without any noticeable roughness; yet, I can barely get to 25 degrees on the lean side while setting the mixture for cruise without encountering objectionable roughness. I don't know whether there is some mechanism that makes it possible to lean the mixture by increasing the manifold pressure without producing roughness, while the same power setting, arrived at by setting manifold pressure first and then leaning the mixture, would be too rough; but I suspect that the difference may be entirely in the focus of my attention. When I am leaning, I am listening for roughness; so I am very aware of the first small change in the timbre of the engine's voice; when I am descending from altitude, on the other hand, my attention is focused elsewhere, and the engine's tone and feel do not preoccupy me. Perhaps the implication is that when leaning, you ought to go right on past peak EGT by, say, 50 degrees, and ignore the roughness, and that it will disappear as soon as you are distracted by other concerns.

One final operational observation will suffice before leaving this topic of efficient flying. It concerns climbing. Contrary to the general belief, it does not take a great deal of fuel to lift an airplane to a high altitude. The fuel required to raise a 3,000-pound airplane to an altitude of 10,000 feet in fifteen minutes is, even at the unfavorable mixture settings used during the climb, only about 1.5 gallons. Furthermore, this energy is stored in the form of altitude, and is recovered when descending. What is irretrievably lost is not the fuel expended in order to climb, but the fuel expended in order to cool the engine during the climb. Considering that one might ordinarily climb at 75 or 80 percent of power with a fuel specific of .55 to .60 pounds/hp per hour, whereas one would cruise at 50 or 70 percent of power with a specific of .40 to .45, it is easy to see that fuel flow during climb will be from 30 to over 100 percent greater than during cruise, whereas forward progress will be somewhat less. The lack of forward progress is not itself a sign of inefficiency, however; as I mentioned before, $V_{L/D}$ is very close to the speed for best rate of climb, and so the airplane is actually flying efficiently, from an aerodynamic point of view, when it is climbing. It is only in mechanical efficiency that it is lacking, and this is entirely due to the requirement for cooling with fuel during the climb.

On flights of long duration, the fuel used during the climb is not extremely significant, since though the fuel flow during climb is high, the time spent climbing is only a small fraction of the length of the trip. There are so many good reasons for cruising at fairly high altitudes that the fuel cost of reaching them remains comparatively unimportant. To the extent that a cruise climb, rather than a best-rate climb, improves cooling and allows you to lean the mixture or clean up the airplane by closing the cowl flaps, it may be desirable in spite of the small departure from the most efficient speed which it entails, and even in spite of the added time it makes you spend climbing.

To make the best use of the energy that you store in the form of altitude, you should make the descent as gradual and long-lasting as you can, and descend not by increasing speed but by reducing power. Power reduction should be accomplished by reducing rpm, if possible, and then readjusting the mixture. A rate of about 500 feet per minute is a good compromise between efficiency and reasonably rapid descent, but the slower the better. As a rule of thumb calculation, divide the distance you have to descend by your rate of descent, and multiply the result by your groundspeed in miles per minute. For example, suppose you are making a 180-knot groundspeed at 17,500 feet, and your destination airport is at 1,000 MSL. Subtracting 2,000, the pattern altitude, from 17,500, you get 15,500; dividing by 500, it will take 31 minutes to make the descent, and you will cover 93 nm in the process. If because of wind your groundspeed is going to change significantly during the descent, you can make an appropriate correction in this figure. One of the adjustments in thinking that turbochargers require is in descent planning. Ninety miles away seems premature to begin a descent; but if you delay it longer, you end up diving for your destination and wasting a little of the climb fuel you could otherwise have recovered.

Efficient flying and its big brother, flying for maximum range, involve a small number of basic elements which were apparent from the Breguet range equation. Speed is one; the ideal speed is $V_{L/D}$, which is a very low speed compared to what you prefer to think of as cruising speeds, but since it is an indicated speed, by climbing to high altitude you can combine good groundspeeds with efficient operations. In general, however, slowing down will improve efficiency.

Engine fuel consumption is another. A lean mixture, as lean as smooth running and the manufacturer's limits permit, is always desirable. When it is possible to lean the mixture by other adjustments in flying procedures, such as cruise-climbing and partially closing cowl flaps (or alternatively climbing at $V_{L/D}$ with cowl flaps open), it should be considered. During the descent, the mixture can usually be left as it was at altitude, and enriched only when power is added.

Airplane weight is a third. As you saw, fuel required to go a given distance increases somewhat out of proportion to the separate elements of efficiency, and the fuel requirement is in the form of a percentage of landing weight. It is more efficient to fly without carrying superfluous weight, which includes excess fuel; in other words, the lighter the weight at which you land, the less fuel you will have used, other things being equal, to make the trip.

There are many strategic considerations in flight planning—choice of altitude, routing, weather, time of day, leg lengths, and so on; absolute efficiency is only one element in a more complex judgment. But knowledge of the principles of efficient flying is not useful only for exceptional long-distance flights; it is important to all flying, and increasingly important as fuel costs rise and as the availability of fuel becomes less certain. In a sense, the entire history of aviation has been like a long flight; the destination is still far off, but we have to throttle back and fly efficiently because the whole world's fuel tank is running dry.

7

RELIABILITY

Flying a light airplane across open ocean is commonly viewed as requiring either exceptional courage or exceptional foolishness. Neither is in fact necessary. The safety record of light airplanes crossing oceans is probably no better or worse than that of light airplanes crossing mountains, deserts, or cities. The lack of any landing places for hundreds or thousands of miles nevertheless seems to give ocean flying a particularly ominous quality, even for pilots who fly at night or in instrument conditions or over rough terrain without any concern. Even when one has subjected the project to rational analysis, there remains an underlying anxiety which may emerge, in mid-flight, in the form of a sudden and scalp-tingling awareness of the tattoo of the engine, or of some nightmarish hallucination produced by fatigue, moonlight, or clouds.

Any long flight, however, be it over land or sea, at home or abroad, involves the same concerns over reliability. One tends, in undertaking long hops, to treat terrain with increasing indifference, and to view weather systems and the time of day as incidental. As I have mentioned before, it is more difficult to avoid weather systems on long flights; and the pressure to minimize the length of flights which already seem overlong makes one opt for direct routes over, say, bodies of water or mountainous areas, where the pilot making a shorter hop might prefer to take his time and follow a road or other detour offering better chances for a safe emergency landing.

Every pilot's principal preoccupation is with his engine, even though other systems in the airplane may at times be equally indispensable. There is something about engine stoppage, which you imagine as sudden and irreversible, that arouses particular anxiety. You allay that anxiety with statistics. It is sometimes reported that aircraft engines fail once in every 50,000 hours of flying—a comforting statistic, were it not entirely without foundation. In fact no method exists for compiling statistics on the reliability of engines, because failures are only reported when they result in injuries, deaths, or significant damage to the airplane. If you talk with an engine repair shop you get quite a different slant on engine reliability than from a Cessna salesman.

Though one imagines an engine failure as a sudden thing, many failures are gradual, and furthermore many forced landings are made with the engine running. What really concerns the flier over water, mountains, desert—over anything, really, except an airport or an empty highway—is not just engine failure, but any malfunction of the airplane that obliges him to land immediately. On this, no statistics exist at all. I myself have, in 3,000 hours of flying, made five "forced landings." One occurred in the course of testing a modified exhaust system, and doesn't really count. One resulted from a magneto failure, and could have been avoided if I had analyzed the problem and simply switched to the good magneto. One was caused by a battery boil-over, which filled the cockpit with poisonous, choking fumes. One resulted when a fitting broke in an oil line, eventually emptying the engine of oil; and one involved a broken crankshaft. If we discount the first two as being atypical and muddleheaded respectively, we are left with three. The oil line problem was the result of poor detail design on my part, and so can be dismissed—though oil lines have broken in factory airplanes, too. Even making these generous corrections in my personal statistics, I have had one forced landing per 1,500 hours of flight. In addition, I have had a lot of problems which could have been more serious had they occurred on a long flight over water; several alternator failures, two vacuum pump failures, a broken piston ring.

Most failures seem to take place in equipment which is of a certain age; vacuum pump and alternator failures, for instance, are most likely to occur after a few hundred hours. Dry vacuum pumps have an uncertain but definitely finite life expectancy, and before going on a hazardous trip it is probably sensible to replace any that have logged more than 600 hours. Alternators are also notoriously cranky devices, inferior to generators in reliability, if superior in lightness. On some engines a generator can be installed to replace an alternator, or a special modification can be made.

Many incipient problems can be detected by careful inspection. Before an overwater trip the cowling should be entirely removed and the engine cleaned and carefully examined, with particular attention to possible areas of chafing; old fuel and oil hoses; kinked, frayed, or unsupported wiring; fuel line connections; and spark plug and ignition harness condition and security. Compression readings should be taken, oil screens checked, and the bores of the cylinders inspected with a grain-of-wheat light inserted through the bottom plug hole and a long-handled mirror inserted through the top. A few scores in the bottoms of the cylinders are normal, but scores elsewhere may indicate broken rings. Compression does not have to be particularly high, but it should be fairly uniform and not excessively changed since the last check. A worn engine with low compression may have a high oil consumption, with the attendant implications for range; but it may otherwise be perfectly reliable.

Catastrophic failures—of crankshafts, connecting rods, pistons, valve stems, and so on—are uncommon, and they are difficult, if not impossible, to forestall by inspection. If you have recently bought an airplane and are uncertain of the condition of an engine, one course is to have the worst-looking cylinder (or the one with the worst compression) removed and overhauled. If it is terrible, then the others are suspect; if it is not bad, then there is some reassurance about the others. In general, though, the most fertile area for inspection is in the peripheral equipment, assuming that the engine has been running normally for one or two hundred hours. An ostensibly healthy engine with all its wires, hoses, and exhausts in good condition and properly secured is about all anyone could ask.

In addition to the engine itself, the fuel supply and electrical systems should be checked. Fuel tank selectors which are difficult to rotate, or whose handles are loose or don't point in quite the direction they should, have to be dismantled and overhauled. Fuel dye stains around fittings and connections should be examined; the flared end of a fuel line can crack off. A couple of in-flight fires have been attributed to cracks or pinhole leaks in aluminum fuel lines, which have been the industry standard for as long as most people can remember. There is no reason for paranoia about fuel lines, but a visual inspection of all connections and supports is advisable.

Quick-drains should be removed from tanks, sumps and gascolators, and their O-rings should be inspected for cracking and for foreign matter that could result in leaks or, in the case of gascolator drains, in air leaking into the fuel system. Something like a stream of air bubbles leaking into a gascolator through an improperly sealed quick-drain is not a dangerous problem, but it will cause the engine to run a little leaner than it should, and perhaps a little roughly; this in turn may cause pilot anxiety, which can lead to an eventual misjudgment, or at the least oblige the pilot to turn back and land. Even the smallest problems may be serious in airplanes, because of the uncertainty with which they are surrounded. It is not possible to make a confident analysis of many problems in flight, sometimes because of their complexity or obscurity, sometimes because the anxiety of the pilot disrupts his thought processes. I had the instructive experience, on the occasion of the magneto failure I mentioned, of seeing how quickly and mechanically I switched into my "forced landing" mode of thinking, without carefully analyzing what might be wrong with the engine. There seemed to be so many possibilities, and so little time in which to consider them, that I almost automatically focused my entire attention on the incipient landing. If I had been over water, it is possible that I would have given more thought to solving the engine problem, because a safe forced landing would have been impossible; but it seems to me now equally probable that I would have switched into the "ditch" mode of thinking and ditched unnecessarily simply because I believed it was inevitable.

Because one's brain becomes unreliable in an emergency, every malfunction, however slight, is dangerous. That is why a careful inspection of even nonessential components is necessary.

Despite my own horrible record, I think that aircraft piston engines in good condition are satisfactorily reliable. It is not necessary to feel that you are doing something extremely daring in setting off over water, or at night, or in bad weather, with a single engine. Statistically, in fact, there seems to be reason to believe that you would be worse off with two engines than with one. The fatal accident rate is higher in twins than in singles. This curious bit of information has been a fertile field for discussion: how can it be so? The unusual explanation is that most engine trouble occurs shortly after takeoff, and most twins perform so badly on one engine, particularly when heavily loaded at the beginning of a flight, that even with one engine still running they are too much for their pilots to handle. It is sometimes suggested that the idea that one engine is still going prevents the pilot from preparing a forced landing, and so the failure of the airplane to remain airborne takes him unawares; in other words, one is more likely to survive a controlled forced landing than an uncontrolled one. Another theory, which I favor but for which I have been able to find little sympathy among pilots and mechanics, is that if maintenance costs are a burden to many airplane owners, they must be particularly onerous to the possessor of a twin; he would consequently feel a desire—not rare among airplane owners—to put off making minor repairs, and his willingness to indulge that desire would be increased by the feeling that since he has two engines, he can take chances with one of them. Thus owners of twins might be prepared to let maintenance slide a little longer than owners of singles. To this add the fact that if there is a certain statistical probability of an engine's failing, the risk of an engine failure is obviously twice as great if you have two engines than if you have but one; and you begin to see why a twin is not necessarily worlds ahead of a single in reliability. None of the ferry pilots I have talked to ever expressed a preference for twins over singles; quite the opposite, in fact. They do not in general consider a mechanical failure a major danger; navigational error, adverse winds, and running out of fuel loom much larger in their thoughts.

Over land, any mechanical trouble involves a certain set of hazards; over water, a different and more serious set. In either case, the most important element is control. The pilot must remain in control of the airplane, try to plan his landing intelligently, and not allow events to overtake him in the final seconds. If sudden engine trouble occurs, the pilot must first satisfy himself that the trouble is geniune, and not some problem of his own making. A few pilots each year crash airplanes simply because they forget to switch to a full fuel tank, or because they pull out the mixture handle when they meant to get the cigarette lighter. It may seem incredible, but such simple errors often elude diagnosis by a frightened pilot. A quick check of cockpit controls does not take long. The engine requires four things in order to run: fuel, air, ignition, and lubrication. If the engine suddenly stops dead, the problem is almost certainly with the fuel supply, so switch to a tank in which there is fuel, turn on the boost pump, look at the fuel pressure gauge, and wait to see what happens. If lines have run dry it may take as long as thirty seconds for the engine to get going again, so one has to be patient. In the meantime, verify that the throttle is forward and that the mixture is where it would be for cruise. It is neither necessary nor advisable to enrich the mixture; if the engine was running before with a mixture leaned for cruise, it will run again. If the engine has stopped entirely there is no point consulting the mag switch. Look at the oil pressure gauge; as long as the prop is windmilling, it should continue to show pressure.

If a forced landing is necessary it will be delayed by slowing to the speed for minimum power, which is quite a bit lower than the best glide speed. The best glide speed is best only in the sense that it yields the flattest glide. If you want to cover a lot of distance to some far-off landing place, you should glide at the best glide speed, which is close to the speed for best rate of climb. The best glide speed may be specified in the pilot's handbook. It will be a little lower than the best range speed.

Engines that begin to run roughly but continue to run may have any of a number of problems, ranging from plug fouling or magneto failure, which can usually be detected by switching to each individual magneto and observing whether the engine runs differently on one than on the other, to mechanical failure in the major moving parts—pistons, valves, pushrods, crankshaft, and so on. An engine may continue to run for some time with its crankshaft in two pieces; but once an engine has started to roughen up, unless you are absolutely certain of what is wrong with it, there is no choice but to start down immediately rather than to wait for it to fail altogether. A forced landing is much safer if it can be made with power. You'd be surprised how difficult it can be to get a fast airplane into a plowed field.

One great bugbear that, like a city-dweller's fear of being dismembered by a madman, rarely materializes in reality, is that of an engine failure in IFR conditions in a single-engine airplane. Over high terrain, there is an excellent chance of the overcast continuing right down to the deck; over low terrain, there is a good chance of breaking out at some height above the ground. It would be nice for the flier to know at all times where he is and what the weather is like down below. IFR at night over mountains—time to kiss your arse good-bye, as wits say; though even here there have been instances where knowing his position on a VFR chart would have made it possible for a pilot to turn toward lower, flatter terrain. This is all very well to say in retrospect, but no one follows his progress on two sets of charts simultaneously. The chance of an engine failure is too remote.

It might be of some interest to set up a minimum rate of descent with the engine at zero thrust—a bit of throttle added to an idle—and with flaps first up and then down, and to see what a landing approach in this configuration would look and feel like. You might find that you could maneuver or flare more readily with a little extra speed, and that the extra rate of descent was not noticeable.

Over water, where accuracy landings are unnecessary, having power available for the landing is not such an important consideration; more important is to stay out of the water, if possible. Engine trouble over water means that you have to find the condition, if any, at which the plane will remain airborne for the longest time, and then devote your efforts to calling for help and looking for surface ships. Once an engine starts running strangely, you must resign yourself to the fact that it may quit at any moment. If, for instance, you drop a valve into a cylinder, there will first be noise and roughness, but the engine will keep going. Eventually the broken valve will probably puncture the piston, dropping metal fragments into the sump and allowing oil to pump out through the piston. The scrap metal may or may not do any harm to the engine (there are screens and filters to stop it) but at any rate the oil will gradually be going overboard. A broken piston ring can be completely undetectable, but it can bring about oil consumption rates of a quart an hour and empty the engine before you finish the flight. The lack of any means to measure the oil quantity in wet-sump engines is regrettable for this reason, but the chance of engine trouble leading to loss of oil is fairly remote.

In fact, the chance of any engine trouble at all is fairly remote. Airplane engines are not delicate; even old, worn, or poorly overhauled ones will run and run, and even the longest overwater trip is not going to require more than fifteen or at the outside twenty hours of operation. If the oil screen is free of metal and the spark plugs are clean when you take off, the chances are good that even if something starts to go wrong during the flight the engine will still be running after 15 hours. Not always, but usually. You can take comfort from the fact that single-engine airplanes are daily delivered across the Atlantic Ocean and the Pacific, and engine failures are virtually unheard-of.

In case of an emergency landing on land, the basic survival techniques are fairly well known. Unless the aircraft is in a position in which it could not be seen from the air, stay with it. Mark the area so as to make it more visible from the air. Ensure that the ELT is working. Erect a shelter from sun or rain. Reconnoiter the area for water; and so on. There are many source books on survival written for different markets: hunters, backpackers, pilots. The *Air Force Survival Manual* covers survival in all parts of the world, and presupposes a group approach, though the downed pilot is just as likely to be alone. Any airplane that makes flights over unpopulated terrain, or flies at night, or in instrument conditions, ought to have on board a basic survival kit, including not only such items as matches, a compass, nylon rope, etc., but also a blanket or sleeping bag, a container of water, and a first-aid kit. The gadgets, such as fishing kits, that survival kits often contain are not of primary importance; warmth, water, and the treatment of injuries are of far greater urgency, assuming that help will come within a few days. Only if you are planning to be lost in the wilderness for a week or more will you find use for the complete Robinson Crusoe tackle that survival kits often contain.

The contents of ocean survival kits for the Atlantic are specified by the Canadian authorities, and commerically available kits usually correspond precisely, at least in their bare outlines, to the Canadian requirements. It is worthwhile to familiarize yourself with the contents of your survival kit; they may be good for a few laughs.

I made several ocean flights with a ''two-man life raft'' and survival kit. Like all emergency rafts for aircraft, this one packed into a very compact container and was to be inflated, in principle almost instantly, by a pull on a lanyard. When I finally decided to inspect the contents of the kit myself, a couple of years had passed since the last relicensing. I suppose it was not particularly surprising, nor to the discredit of the product, that the raft failed to inflate on the pull of the lanyard. I laboriously inflated it by hand, using a bellows supplied with the raft. I then discovered that though it was possible for two persons to sit on the raft, it was only barely so; their legs had to be either uncomfortably folded beneath them or dangling over the sides; and the freeboard was no more than a few inches. The test took place in a swimming pool, yet the raft was nearly swamped by the enthusiasm of a few nearby children. Furthermore, we had to sit in the bottom of the raft, with only a layer of rubberized fabric between our buttocks and the water; not bad in a Los Angeles backyard, but much worse, I imagine, somewhere off the Aleutians. To sit on the flotation ring for relief from the cold would have pushed it nearly under water. It seemed to me quite obvious that while this raft could, in an emergency, keep two persons (even two large men) afloat, it would have delayed their demise only a little.

I happened also to have something called a four-to-six-man life raft. This inflated when its lanyard was pulled, though it too had been packed several years before. It was much larger and sturdier. One could easily sit on the edge and move about the raft without any danger of capsizing it. It had an inflatable arch and a canopy that would serve as a sail and a sunscreen. Folded up, however, the four-to-six-man raft was barely larger than the two-man raft, and not much heavier. Certainly if there had been any thought in my mind of actually ditching, I would not have considered using the two-man raft, whereas I might have entertained the thought of a few days on a tropical sea with a lady friend in the four-to-six-man raft with some Walter Mittyesque pleasure.

There appears to be a principle in survival equipment, as in frozen vegetables, that the advertised quantity must at least be halved to be realistic. Thus, if a package of frozen peas says, "serves four to six," any homemaker knows that it will barely satisfy two, unless it is so distasteful that no one will eat it at all. Similarly, a two-man raft might be used as a kind of sidecar to save a dog from drowning; a four-to-six-man raft will possibly handle two people; and I suppose that if you filled up a four-seater to fly to Europe, you would need a ten-man raft, at least, if you had to switch to surface transport along the way.

Different makes of rafts may be more or less misleadingly labeled. One four-man raft of which I have seen pictures (in action) could support four people in calm water, so long as none of them actually climbed into the raft.

In addition to my disappointment in discovering that my "two-man life raft" was barely adequate to support one man and a lissome teenaged girl, I was a little startled to discover that "two days' supply of concentrated food, or its equivalent" consisted of two packages of hard candy resembling Life Savers. This must have been the "equivalent," since it was certainly not what I would have called food.

A problem with rafts, apart from their size, is that they can be quite difficult to use. One does not, as a rule, simply land on the water, step out onto the wing, inflate the raft, and step aboard. A ditching is an emergency; it is fraught with uncertainties. High-wing airplanes obviously provide no reliable foothold for the departing pilot, though even ones with fixed landing gear can be ditched satisfactorily. Time of flotation varies widely; some

One man (top) sits in a "four-man life raft." Three other men attempt to join him (middle). The raft sinks (bottom). This took place in calm seas.

airplanes have gone down in a minute or two, others have floated for days. Usually, because the engine cowling doesn't hold air very well and the engine is a large concentrated weight, an airplane settles in a nose-down attitude after ditching. If the airplane has a low wing, the cabin will be slow to fill with water, unless the ditching takes place very early in the flight when all fuel tanks are still nearly full. A high wing is less convenient; the cabin begins to fill with water, and the pilot has to get out, raft in hand, when he is already partly under water. So while it may be remotely possible to get from a ditched low-wing airplane into a raft without getting wet, it is impossible to do so from a high-wing plane.

The difficulty of getting from the plane into the raft, and the amount of time available in which to do it, are important; in a crisis, things are forgotten and get left behind—flashlight, ELT, what have you. If the plane is sinking fast, even the raft may be forgotten.

Getting wet is not merely an inconvenience. In northern oceans the water temperature is so low that a person floating in the water will not live more than a few minutes. Even if he only got wet and then hauled himself onto the raft, he would be liable to serious harm from exposure. There are various exotic solutions to this problem, including survival wet suits, but these are of limited utility unless they are worn, and a person cannot wear a wet suit continuously against the moment he may have to use it. Use of a wet suit presupposes a certain type of ditching, with ample prior notice.

It is all really too fanciful. You carry certain basic pieces of survival equipment because it seems foolhardy to go over water with no protection at all. But it is impossible to protect yourself against every remote eventuality. The majority of airplanes that ditch in the ocean disappear with their pilots; the few pilots that are saved are those that ditch alongside ships, or in otherwise impeccable conditions. What is more to the point is that ditchings are very rare. I do not know where one would go for precise figures (hardly anyone has much interest in publicizing them, though I am told the FAA in Hawaii has complained about the amount of money spent annually on search and rescue operations for amateur ocean pilots), or what one would gain by knowing them. In the final analysis, when you make an ocean flight you are risking your life; you must simply face that fact. The risk is small; whether survival equipment of exotic types would make it any smaller is impossible to say.

To sum up the basic requirements for survival over land, they would be warm clothing, a blanket or a sleeping bag; matches; water; and a flashlight. These four items would answer to basic and immediate needs. First aid for bleeding and some fractures can be fashioned, usually from clothing and debris. One can go without food for a long time, but dehydration and hypothermia—chill—are serious dangers from the beginning. The flashlight, apart from being an important piece of backup equipment in the cockpit, is necessary at night to make use of the other things. All additional equipment is gravy.

At sea, because the possibility of long exposure is greater, additional equipment is needed. But while people are regularly rescued from crashed airplanes on land, mostly because help arrives soon, at sea the chances of prompt rescue are slimmer, and the difficulties of survival for long periods are much greater.

In thinking about survival, I always come back to the idea that there are some people who are simply not going to want to fly across the ocean, and there are others who will. Those who don't want to will not be persuaded by the presence of a life raft, no matter how well equipped; and those who do will not be deterred by the absence of one. For me, it was sufficient reassurance that many people had done it, and few had come to grief. In retrospect, the satisfactions and pleasures of my ocean flights have certainly warranted the risks, whatever they may have been.

8

MEXICO AND SOUTH AMERICA

The Latin American countries provide a pilot wishing to get off the beaten track with a myriad of destinations that will challenge him without exposing him to long overwater legs or to the ruinous price of fuel in the North Atlantic, Europe, or the Pacific Islands.

Mexico is familiar enough. Many California pilots make frequent excursions to Baja California, where airstrips of every kind give easy access to excellent fishing, beautiful and empty beaches, camping, hiking, and a number of very cheap and very expensive hotels. Some Americans feel timid about flying in Mexico, because they have heard that the Mexican authorities often impound airplanes, and even their owners, for no reason at all.

Recovering impounded airplanes has become a lucrative business. The risk of visits to Mexico is held to be so great that many FBOs within flying distance of the Mexican border explicitly exclude Mexico as a destination for their airplanes.

Behind all this is the dope trade, as well as a certain amount of xenophobia and racism on both sides. Most of the airplanes that have been impounded in Mexico were impounded because they were carrying dope, or were suspected of carrying it. In some cases airplanes have been impounded because of other violations of regulations. But cases of completely arbitrary and unjustified seizure of aircraft are as uncommon in Mexico, I suspect, as they are in the United States. Most pilots flying to Mexico find nothing to complain of in the treatment they receive.

All Latin American countries, including Mexico, require advance notice of overflights and landings. In most cases twenty-four-hour notice is required; sometimes the number is forty-eight hours. The requirement is apparently not enforced in Mexico for tourist flights arriving from the United States and returning to the United States. For such trips, no visa is required; a "tourist card" is issued at the border and reclaimed when visitors leave Mexico. For flights in transit to or from other countries, a "transit visa" is required and can be obtained at a Mexican embassy or consulate in the United States. A general declaration is required, but rarely requested.

Most insurance policies exclude foreign flights, and a special policy must be obtained for flying in Mexico or Central or South America. One firm specializing in this type of insurance is MacAfee and Edwards, 3105 Wilshire Boulevard, Los Angeles, California 90010. The cost of insurance is small; usually about $15 for a four-day trip. I paid $130 for a month-long policy for all of Latin America in 1980. All countries require insurance; some ask to look at your policy, or some evidence of insurance, as you cross the border.

Latins have a passion for paper work, rubber stamps, and fancy signatures. They also like to see paper work done neatly, with flair, and preferably on a typewriter. Officials who can barely read a newspaper have perfected elaborate signatures combining a completely illegible scrawl with various triangles, lines, and ellipses worthy of an eighteenth-century prime minister. It is extremely important, on arriving in any Latin American country, to be armed with every possible paper, authorization, license, and statement, and for as many as possible of these to contain one or two stamps, seals, imprints, and signatures, preferably with flourishes. Copies of telegrams or letters sent to notify aviation authorities of your impending arrival (the cable addresses are in the *International Flight Information Manual* and in Jeppesen trip kits) must be carried; these too must have seals and signatures or at least some kind of scribble on them. If the telegraph company does not scrawl a signature across a corner of the copy, scrawl one yourself. If you ever run across a press which is used to emboss a seal onto paper—public libraries used to use them to mark books—you might try to buy it cheaply; an embossed seal would probably work magic.

All Americans visiting Mexico are surprised by the amount of useless paper work generated by each flight. Every flight requires a flight plan, and if this cannot be filed because there is no aeronautical authority to accept it at the airport of departure, then a flight plan is made up retroactively after you land. I can't imagine the purpose of creating flight plans after the flight has ended, unless it is to keep tabs on all the movements of airplanes, again because of the drug trade. You would think that a smuggler would have little trouble beating the system. At any rate, extravagant as the paper work in Mexico may seem, the farther south you go, the worse it gets. Most South American countries require not only flight plans, but also government authorizations for every flight. These are easy enough to get in advance by notifying the civil aviation authorities of your intended route, all proposed landing places, and the approximate dates you expect to reach them. In the case of a country like Colombia, where two-hour notice by means of a flight plan is acceptable for entry into the country, the flight plan should include the entire itinerary, dates, and so on. Prior notice by telegram or letter is better, however; flight plans can go astray.

Mexico does not permit night single-engine flights; neither does Colombia. Some Latin countries do; the *IFIM* does not go into detail about these regulations, and if you want to find out about them you should send an inquiry along with your request for authorizations.

You will note in the *IFIM* that different countries word their requirements differently, some requiring advance notice of a flight, others insisting upon authorization—that is, upon your having a reply. A few require that you actually receive the reply before you enter the country; others exempt private airplanes on tourist flights from that requirement, and say that if you have not received a refusal, you may enter. The countries that require authorizations, however, require that you provide for a prepaid reply—fifteen words is the usual minimum—when you request your permissions.

In general, countries make little attempt to make their regulations and requirements available to foreign pilots, but they act surprised when you fail to comply with them. One annoying condition, in my experience, is the complete lack of advice for the arriving foreign pilot. You pull into a big international airport like that in Bogotá, Colombia, and you have to find your own way to the half-dozen or so offices which need to know of your arrival. No one is available to direct you, and no one has thought of the simple expedient of a mimeographed sheet with a list of required stops on one side and a map of the airport offices on the other.

Customs and immigration officials are undeservedly notorious for attempts to soak you for a little money on the side, either by looking at your paper work with such a doubtful expression that you hasten to offer them bribes, or else by inventing bogus charges that you are afraid not to pay. Such shakedowns are inevitable, but infrequent. You can be easily intimidated by these little men in their brown uniforms, with pistols at their hips; it may help to ask for a detailed explanation of every charge, and to request receipts and make sure the receipt is for the same amount as you have been asked to fork over—it sometimes isn't. If, for instance, somebody is hitting you up for an overtime charge at two in the afternoon on Monday, you are well advised to delay; claim to have no money handy and go to some other office. It is sometimes good to drag a friendly official around with you; some are more hesitant to cheat you in the presence of others.

However, you have to keep in mind that many charges unheard-of in the United States are normal in Latin America. You may pay a fee for an approach, for landing, for parking, and even for use of en route radio facilities, or a kilometrage fee for all the flying you do within a country. These are all perfectly legitimate charges, and not attempts by petty officials to line their pockets (or, putting it more kindly, to be able to bring home a toy for their children). I have found that in general, one should be suspicious of people who say flatly, after a few moments of reflection, "Twenty dollars," or "Fifty dollars," and do not volunteer a receipt. Anyone who presents an itemized bill for a non-integer amount, however weird the charges listed, is probably dealing fairly.

You must also remember that you are, if not absolutely then relatively, a wealthy person traveling in an extravagant and expensive way among much less wealthy people who as long as they live will never be able to do what you are doing. It is natural that they should be tempted to extort a little money from you, and to comply with their demands will often smooth your way at a cost that is small compared with the cost of gasoline or even of all those approaches, radio contacts, and kilometers. The main thing, really, is that you probably would rather distribute the $50 among fifty beggars and bootblacks than hand it to one pistol-packing border guard; but you can't always have your way.

The volatile political conditions in Latin American countries may limit the choice of stopping points; unless a civil war is at its climax, however, you are unlikely to encounter serious difficulties even in countries which are under martial law or where guerrilla fighting is common. You may not enjoy yourself much, however; fear of kidnapping, or of the demolition of your airplane—a symbol of North American exploitation—can marvelously reduce your enjoyment of food and sight-seeing. When I traveled down as far as Chile in 1980, I thought that Costa Rica was the only place in Central

America where you could comfortably stop (Panama was possible, but unnecessary). However, I greatly enjoyed a previous trip to Tikal and other destinations in Guatemala, and would recommend them to anyone willing to risk the hazards of a visit to a war zone. The hazards are not that great, and in spite of everything tourism continues.

The weather in the tropics changes little year round and is characterized by clear mornings with the rapid buildup of cumulonimbus toward midday and heavy thundershowers in the afternoons. This pattern is largely confined to the land, and one can fly down the coast, or at the most, several miles offshore, all day long, unless the wind is carrying the storms out to sea. Tropical thunderstorms are said to be less violent than those in temperate zones, but they are violent enough to be worth avoiding. However, ground controllers don't have radar for weather depiction down there, and if you have to fly through a thunderstorm to reach your destination then you just have to take your chances. I have flown through a couple, and they seemed quite violent enough.

As you emerge from the tropics to the south, you get into weather conditions reminiscent of those in the northern temperate zones, except that the seasons are reversed—winter begins in June—and the prevailing wind flows are, too. For instance, the western coast of Peru is a desert, receiving little or no rain; the rain falls on the east slopes of the Andes—the opposite of the pattern in California. However, a cold ocean current sweeps the Peruvian littoral, and during much of the year a thick stratus layer hangs on the coast; it is centered over Lima, which rarely gets any rain, but rarely gets any sun either. During the winter the air over the Andes is relatively stable; usually there are few clouds in the morning, and a thin layer of stratocumulus with scattered cunims appears as the day wears on. There is little turbulence over the mountains in the morning. Summer weather (December), I am told, is stormier throughout the continent, but I have not experienced it.

The Andes are the dominating feature of the western part of the continent. For much of their length they consist of a narrow, high plateau, from which numerous peaks rise to altitudes approaching 23,000 feet. In most places they can be comfortably crossed at 18,000 feet, and in a few places at 15,000. They are so narrow—100 miles or so, in many places—that the crossing does not tax your navigational abilities; however, a turbocharged airplane is a great advantage for a pilot who is not intimate with the passes and the weather conditions. Sheer altitude is a great advantage in rugged terrain. In some places the Andes consist of jagged, snowcapped peaks alternating with chasms miles deep—for instance, in the vicinity of Machupicchu; but for the most part the valleys between the peaks are wide and flat, and often they contain water or signs of human habitation. In other words, a forced landing in the Andes is not as desperate a prospect as readers of *Alive!* were led to think. Of course, the farther south you go the colder is the air at the mountaintops. In Ecuador and Peru only a few peaks are cloaked with permanent snows; in central Chile snow is more prevalent, but the chain consists of no more than a single narrow ridge. If the visibility is good and you can climb to sufficient altitude, crossing it is easy, provided the wind is not strong.

Beyond the mountains to the east lie, in the north, the Brazilian jungle, and in the south the Argentine pampas, or grasslands. The grasslands resemble those of the central United States; the Brazilian jungle, however, is more like the ocean, except that it is made of trees. It is nearly trackless—occasional prominent rivers provide the only landmarks—and predominantly uninhabited. The same considerations of navigation and survival apply to it as apply to bodies of water; difficult as it may be for readers of Tarzan books to believe, a pilot down in the jungle, even if he survives the landing in the treetops, is unlikely to fare so well in the jungle below, because of the peculiar blend of insects, waterborne diseases, and the difficulty of traveling through the extremely dense undergrowth. Because of the inhospitable nature of the South American interior (except in the south), most flying is done around the perimeter of the continent, and that is where most of the radio facilities are located.

Radio facilities are more numerous than you might expect; there are VORs on spacings of 200 nm or less along many routes, and quite a few DMEs. All are said to operate only sporadically; it is wise to check NOTAMs (Notices to Airmen) before departing on trips whose navigation depends heavily on one or two beacons. NDBs are numerous, but not especially powerful; in my experience, most en route NDBs have about the same range as a VOR. En route communication is theoretically handled (in English, Spanish, and Portuguese) through single frequencies, such as 126.7 or 126.9 kHz; but it is rarely possible to get an answer on these frequencies, and one usually has better luck communicating with towers along the route. They accept VFR and IFR position reports. Most towers have frequencies in the 118s; there are few approach control stations and few ground control frequencies, and even fewer radar controllers.

For international flights and for some internal flights, the ICAO (International Civil Aviation Organization) flight plan form is used. Whether the flight is IFR or VFR, it is treated by en route radio facilities as though it were IFR, and estimates are frequently requested. They are a nuisance, and it is easier simply not to contact the en route facilities, unless you are concerned about making sure people know where you are in case you go down. Whether anyone would actually go out looking for you in much of Latin America if you were to disappear, I don't know. On the other hand, the virtual interchangeability of VFR and IFR is sometimes a convenience; if you want to pick up an IFR clearance en route, it's easily done. The traffic density, except in the busy terminal areas, is very low, and flying in cloud without a clearance is common enough, particularly since most cloud build-ups are of limited extent; you go in one side and are soon out the other. Given the paucity of traffic, the extreme diligence of local controllers at major terminals is incongruous. For example, at Lima a VFR flight often has to begin with an IFR climb to VFR on top. The clearance always includes a limit—Salinas to the north, Asia to the south—50 miles or so out; even if you top the overcast at 3,000 feet 5 miles from Lima, you can't cancel and proceed visually as you would in the United States; you have to go to the fix at which your clearance ends, even though it may take you far out of your way. The temptation to proceed visually and make up some suitable position reports to satisfy the controller is very strong.

Many of the officials at whose mercy you find yourself are not themselves pilots. The colonel in Lima whose job it is to issue or deny authorizations to foreign aircraft to visit Andean towns like Cuzco, an indispensable tourist destination in Peru, lives in a fantasy about the difficulty of such flights. He imagines that Cuzco lies in a tunnel, and that only a virtuoso pilot can land there. By the same token, the suggestion that you might want to fly from Lima to Bogotá along the east side of the Andes—that is, over the jungle—is greeted with amazement in the Lima briefing office, even though the inland route is shorter, has (in winter) a lesser likelihood of thunderstorms, and avoids the bureaucratic complication of overflying Ecuador. While it is always sensible to seek the advice of local people about weather and route conditions, it is important to distinguish between the fevered imaginings of nonpilots and the real experience of practicing pilots. It is also important to distinguish the requirements and liabilites of different types of equipment. The advice of a pilot who negotiates the valleys in a Cherokee 180 is of little help to the operator of a turbocharged twin; and vice versa.

Despite the remoteness of some locations, mechanical services in Latin America can be very good—better, in fact, than the run of the mill in the United States. Mechanics are hardworking and resourceful, as they must be, since parts are hard to come by. Parts are dear, but shop time is comparatively cheap. Fuel, too, in Central America and in Mexico, is generally less expensive than in the United States, gradually becoming more costly as you go farther south. Prices vary greatly by country; in Ecuador auto gas was $.20 a gallon in 1980, and avgas was $1.00; next door in Peru, auto gas was $.80 a gallon, but avgas was $3.50. The acceptable modes of payment also vary; Mexico accepts credit cards such as Visa or MasterCard; farther south credit cards diminish in usefulness, but traveler's checks are accepted. Still farther along, only American dollars are accepted—not even the local currency. One of the most useful things you can take along on a trip to South America is a thick wad of dollars, including small denominations, since the inability to make change, real or pretended, is endemic.

Besides the usual papers required of a traveler, South American countries show a particular interest in the "General Declaration" (Customs Form 7507, available in pads from the Department of the Treasury), which usually has to be presented in quadruplicate. The information required is rudimentary, however: identification of the flight, names of pilot and passengers, points of origin and destination. Once stamped by the appropriate authorities, the pilot's copy of the General Dec is more valuable than his passport, and should be guarded carefully and kept available at all times.

Despite the legendary bureaucracy, which is manageable if one is mentally and materially prepared for it, South America is a wonderful place to travel. It has spectacular scenery, historic structures and places, ancient lost cities, untouched wilderness, a temperate climate, and various excellent cuisines. The flying conditions are good. And South America provides North Americans with their only opportunity of visiting another continent without crossing a body of water—an important qualification for those who love adventure more than risk.

9

SETTING RECORDS

Since the early years of aviation, pilots have had a fascination with records. The Paris-based Fédération Aéronautique Internationale (FAI) has set itself up as the official arbiter of record claims, and issues a *Diplôme de Record* to record holders. At first everyone's concern was with going farthest, highest, and fastest; but as aviation grew and diversified the numbers of types of records increased, and a swarm of categories and sub-categories came into being, in recognition of the fact that while the best performance, for instance, of a light airplane may not be equal to that of a heavy jet, still each is exceptional on its own terms. At the same time, certain records, like absolute straight line and closed-circuit distance records, remain open to aircraft of all types.

In the United States, the National Aeronautic Association (NAA) serves as the arm of the FAI, handling preparation and verification of record attempts, submitting the results to FAI for recognition, and recognizing, on its own, a special category of national records.

To set a record, one would first examine the existing records, published in a 150-page three-ring binder by NAA, a subscription to which costs $35 annually. Revisions are supplied as they are required—which is quite often, because in certain categories the record business is a lively one. Having decided what record to surpass (and it is necessary to surpass existing records by a certain margin, in most categories 1 percent, but in some, 3) one would contact the NAA in Washington (821 Fifteenth Street N.W., Washington, D.C. 20005) and request the necessary application forms for a record attempt sanction and also for the FAI Sporting License which is required of all record seekers. (This "sporting license" is a meaningless slip of paper—like most licenses—which merely enables the issuing institution to collect money from the license holder.)

A sanction is the official acknowledgment of one's intention to attempt a record. It is of 90 days' duration, and may be extended to 120 days on request. Sanctions for a given record are issued to only one applicant at a time, and may be issued at any time more than 30 days in advance of the intended period of the sanction. Specific arrangements for the attempt are made by telephone or correspondence, or, if it is convenient, by meetings with NAA representatives.

A sanction costs money. There is an elaborate schedule of fees which the NAA supplies in its Record Attempt Kit; sanction fees run from $925 for a world absolute record to as little as $100 for "additional" class records for light aircraft under 1,000 kilograms, in takeoff weight. "Additional" records are ones claimed in addition to some basic record, and on the same flight; thus, if you were to attempt a closed-circuit flight from Los Angeles to Las Vegas and return, you might also apply for an additional record for one leg of the flight (or both individually).

Certain formalities must be accomplished, including a weighing of the aircraft to establish its class, and in some instances the installation of a barograph, which traces the altitude during a flight and proves that no clandestine landings were made. For records over established courses and between geographical points, FAA radar or tower controllers can provide start and finish times; "official" NAA observers are not usually necessary.

After a record attempt has been made, the documentation is turned over to NAA along with a "registration fee," which is somewhat greater than the sanction fee, ranging from $3,000 for world absolute records to $150 for very light aircraft. (All these fees are discounted for members of the NAA; the $20 annual membership fee is obviously a good investment for anyone intending to make a record attempt.)

Because the fees and miscellaneous charges for even modest records run into hundreds of dollars, and may mount to thousands if attempts on several records are made, some pilots obtain sponsorship for their attempts, either from well-heeled private enthusiasts or from commercial enterprises.

The taxonomy of records is bewildering. The basic categories of speed, altitude, payload, and distance are cross-pollenated to produce hybrid records—various weights lifted to various altitudes, for instance. Some records, such as the largest number of astronauts simultaneously outside a space vehicle (currently standing at two) seem to have been invented for a special occasion. As far as general aviation pilots are concerned, most of the world absolute records for aircraft are out of reach (2,193.16 mph, 314,750 feet altitude, and so on) though those for straight-line or closed-circuit distance (12,532.28 miles and 11,336.92 miles respectively) could be broken by a specially built lightplane; one is now being developed to do so. The categories that are most fertile for general aviation pilots and aircraft are the so-called Group 1 (piston engine) Class C-1.a, b, c, d, and e. The lower case letters refer to the takeoff weight categories, which are as follows:

a. under 500 kilograms b. 500–1,000 kilograms c. 1,000–1,750 kilograms d. 1,750–3,000 kilograms e. 3,000–6,000 kilograms.

A weight right on a boundary, like 1,000 kilograms, falls into the higher class (C-1.c). (A kilogram is 2.2046 pounds.) In the case of unmodified aircraft, the manufacturer's published gross weight is accepted as a takeoff weight; overloaded aircraft must be specially weighed. Thus, the Piper Twin Comanche in which Max Conrad flew 7,878.17 miles for Cape Town, South Africa, to St. Petersburg, Florida, set its record in class C-1.e, not C-1.d, where its certificated gross weight would have placed it.

In these classes there are absolute distance, speed, time to climb, and altitude records. Another handy category is that of FAI course records. Most absolute course records are out of reach, having been set by jets, but a few belong to light aircraft, and a few exist only because no one has bothered to replace them. For instance, the 1964 record of 33.818 mph between Reykjavik and New York is undoubtedly surpassed every week; but no one has bothered to sanction and register a better time.

In addition to FAI absolute course records there are course class records, again in the five weight classes mentioned above. Since the list of "important" cities between which records can be set is incredibly long, and includes such world capitals as Ogden, Utah, and Minot, North Dakota, all kinds of records are available here. Most city pairs have not even been tried. This is virgin territory. In some cases the speeds are low—for example, 45.93 mph from New Delhi to Sacramento in a Cessna 210—but that is because several stops were necessary en route. On the other hand, I can think of no explanation, short of engine trouble, for the 99.71 mph speed set by G. R. Wood in 1976 between Mobile, Alabama, and Austin, Texas, in a turbocharged Bonanza.

Obviously there are records and records. Some are set by airplanes with special capabilities; these are hard to surpass, and do not necessarily represent a personal accomplishment for the record holder. One SR-71 pilot can fly as fast as another. Other records do not represent a remarkable performance for an airplane, but for a person, such as Don Taylor's round the world flight in his homebuilt Thorp T-18. It took him two months, and his average speed was barely more than that of a running man, but he hand-flew the little airplane, scantily equipped, over ocean and forest, braving bureaucrat and monsoon, all the way from Oshkosh, Wisconsin, to Oshkosh, Wisconsin.

Other records are easy marks; they would not be hard to surpass, and they exist only because no one has bothered to surpass them.

Finally, there are the cheap shots: course records between city pairs for which no record exists. It is not only obvious that a record can be set, but there is no element of competition at all. One might set a record between, say, Los Angeles and Long Beach, or New York and Newark, or San Francisco and Oakland in a Piper Cub, just for the sake of the *Diplôme de Record* to hang on one's wall.

In addition to one-way records there are round-trip records, and these are even easier to set in the "cheap shot" category because attempts are less numerous. Generally, however, it is more difficult to establish a remarkable speed or distance on a round-trip or closed-circuit flight than on a stright-line flight, because of the element of wind. If one wished to set a class straight line distance record in C-1.a to surpass the mark of 1,835.46 miles set by Edgar Lesher in his homebuilt 100-hp *Teal,* one could do worse than to install a cabin tank in a VariEze and take off from Los Angeles during the winter months at a time when a big low is sitting over Utah or Colorado, and a high over Texas. The range of the airplane in still air would be over 2,000 miles, and a tailwind component of 20 or 30 knots might push you all the way to the East Coast. All stright line distance and speed records must be evaluated with that tactical consideration in mind: they do not necessarily represent the performance of the airplane, but rather the performance of the airplane combined with a judicious choice of winds.

Attention to meteorology can help on a closed-circuit or round-trip flight as well. An attempt at a very long flight might, for instance, entirely circumnavigate a low pressure center counterclockwise, keeping the wind behind it all the way. Sometimes the wind gradient with altitude can be put to use, as well as the difference in direction between the lower and upper level winds.

Whenever a change of altitude is to be made on a round trip, the high altitude (downwind) leg should be flown first, since the starting time will be taken with the airplane already at altitude, and it will be able to gain speed on the descent.

Remember that the average speed for a round trip with wind is equal to the total distance divided by the sum of the outbound time and the inbound time. Since the airplane will take more time to fly the upwind leg than the downwind, the speed for the upwind leg will apply for the majority of the flight, and thus the average speed will be lower than the simple arithmetic average of the upwind and downwind groundspeeds.

While absolute speeds over short courses, such as the 3 kilometer course, must be set within limits of altitude, most other records do not involve constraints on altitude. Thus, even on a calm day one could improve a speed over a comparatively short distance by starting the flight at a high altitude, and gradually descending throughout the flight, converting height into extra speed.

A turbocharged airplane, whose speed improves with altitude, actually does not profit from wind gradient on a closed circuit so much as would a normally aspirated airplane, because the turbocharged airplane would always do better on a calm day by climbing to its critical altitude and staying there. A normally aspirated airplane, on the other hand, which in calm air has little to gain from increased altitude beyond 7,000 feet or so, may considerably improve its performance by climbing into a region of stronger winds for the downwind portion of the flight.

Geography can also play a role in record setting, as in general flight planning. When William Windover flew a 75-hp 1941 Interstate Cadet, hardly a high performance airplane, to 30,798 feet in 1972, he did so in April at Colorado Springs, a gliding center where the mountain wave readily carries airplanes with no engines at all to similar altitudes. One could take advantage of ridge lift with a westerly flow over the Sierra Nevada, or an easterly flow over the Rockies, to achieve very high speeds or very low fuel consumptions on north–south legs.

Seasonal variations in pressure patterns can be put to use. A pilot seeking a distance or speed mark in, say, a Mooney 231 would find that the absolute class records were all in the hands of the late Max Conrad, but that he might set a course record from San Francisco to New York, or something of the sort, in wintertime, whereas in summer he might be better advised to fly from New Orleans to Los Angeles. The only way to mitigate the unsportsmanlike aspect of most course records—that is, the fact that there is no competition, and that any speed will do—is to turn in an exceptional performance which, to the astute student of the record book, will have the quality of an absolute class record, even though it isn't one. For example, the 302 mph speed record of a 231 between San Francisco and Washington, D.C., is quite a performance. So is Judy Wagner's 205.23 mph in a normally aspirated Bonanza between Los Angeles and New York, particularly when one notes that the flight was made in June and was followed, three days later, by a 192.41 mph return.

Range plays an important part in the longer course records, because fueling stops cut one's average speed badly. An airplane that can make the course nonstop may make a better time than a considerably faster airplane that must refuel. In this regard, very fast times in light aircraft over very long distances are remarkable performances, because they represent a terrific sustained effort on the part of the pilot; the smaller and less comfortable the airplane, the more remarkable they are.

Many records in C-1.a and C-1.b are held by homebuilt airplanes, and many more could be if someone took the trouble. In fact, I would venture to say that there is not a single record for altitude, speed, or distance in those categories which could not be surpassed by one or another homebuilt aircraft. The Experimental Aircraft Association might augment its already considerable prestige by diverting some of its funds to sponsorship of promising record attempts; the high cost of records is unfortunately a deterrent to many, and it prevents the record book from being a truly representative portrait of the capabilities of modern airplanes.

Even though many world records are quite meaningless, there is something seductive about the idea of setting one, just for the hell of it. Some people have made a hobby of setting records and thinking about setting other ones; and here and there a pilot is scheming to recapture a record which he lost to another, who will in turn scheme to get it back. The same spirit which seeks to increase the number of undergraduates in a phone booth or the number of crawdads consumed at one sitting also yearns to get from Minot to Ogden faster than anyone else has ever done so.

The challenge is to set a record which will be hard to replace; to stay in the books for a long time, in other words, and not to be dumped at the next revision. Once your record has been beaten, it is naught; though you could always print on your business cards, "Former Holder of the World Record for . . ." Some records are very old; and some, given their age, are, like a few old people, in surprisingly good health. The speed of almost 441 mph set by a pontoon-equipped single-engine seaplane in 1934 will probably stand forever, as may the 1928 flight of the *Graf Zeppelin* from Lakehurst, New Jersey, to Friedrichshafen, Germany (though in the next year the *Graf Zeppelin* made even longer flights; but no official record was applied for). It is unlikely that another private individual will challenge the speed of 988.26 mph, made over a 3-kilometer course almost at ground level, by Darryl Greenamyer, who assembled his modified Lockheed F-104 from spare parts. And will the 230-mile flight of a model airplane powered by a rubber band, made in the USSR in 1964, ever be surpassed?

For $35 a year, you can find out. But beware; there is a certain risk of addiction.

ENVOI

There are a few sources of detailed information for pilots planning ocean flights. The *International Flight Information Manual (IFIM),* obtainable from the Government Printing Office on an annual subscription basis, is the basic compendium of foreign requirements and regulations, but it is incomplete (limitations on single-engine night or IFR flight in some countries, for instance, are not mentioned) and provides no helpful commentary at all.

Jeppesen Trip Kits contain, in addition to all the information found in the *IFIM,* glossaries of foreign terms, guides to understanding weather reports and the ICAO flight plan form, and so on. They are indispensable for foreign travel, in my opinion, both because of the copious background information they contain and because they provide a complete and convenient source of navaid information, tower and en route frequencies, and so on, in a form which I, at least, find more convenient than that of the DOD en route supplements.

The Aircraft Owners and Pilots Association (AOPA) provides a trip-planning service to its members and publishes booklets containing much indispensable information on international flying. It also supplies preplanned flight logs for some commonly flown routes, though the one that I had with me when I first crossed the Atlantic was misleading, since it gave waypoints in fractions of degrees, and Gander Oceanic would not accept them.

Canada publishes booklets on Canadian regulations and practices; their Department of Transportation in Ottowa will supply the requirements for survival and radio equipment and so on. Most European countries publish booklets outlining information useful to visiting pilots; these are available through tourist offices, which also provide information about visa requirements.

Louise Sacchi's *Ocean Flying* gives a good deal of information about the heavily traveled ferry routes across the Pacific and Atlantic, which are also the ones that ocean-hopping pleasure pilots are most likely to use.

The U.S. Customs Service will provide pads of "General Declarations," which are the basic customs document that will be required by every country. If two of you travel, it is best to list two crew members and no passengers.

Some countries may require that you have a "ground handling agent." This means a local FBO or other established authority whom you engage to provide you with services during your stay. Getting hold of a ground handling agent can be tricky, and it is one of the reasons for starting your letter-writing early. One way to start is to write to the airport authority, or the aeronautic authority, for the place where you intend to land, and explain the problem.

The world political climate is subject to change without notice, but it seems as I write this that citizens of many foreign nations, especially Third World countries, are, if anything, less fond of Americans than they ever were. It is probably getting less and less practical to gad about the world in an airplane. One homebuilt going from Australia to England was forced down by MIGs in Syria; I'm sure he won't be the last. Costs are going up, and political barriers are growing more numerous; Hawaii and Europe are still accessible, but places in Africa, Asia, and South America may be less and less so. If you ever want to do it, the time to start is now.

INDEX

AUTHOR BIOGRAPHY

Peter Garrison has been a pilot for twenty years and a contributor to *Flying* magazine for ten. An honors graduate of Harvard with a major in English, Garrison began working on a design for his own airplane while still an undergraduate, and finally flew it ten years later, in 1973. That airplane, *Melmoth,* has carried him across both the Atlantic and the Pacific, and logged over 300,000 miles of flying. A free-lance writer, Garrison lives in Los Angeles.

Robert B. Parke fought in World War II as a B-17 pilot. He has thirty-three missions to his credit and a Distinguished Flying Cross. He was called back to duty in the Korean War, where he flew multi-engine cargo planes. He has logged over 6,000 flying hours. In 1952 he joined *Flying* magazine, the private pilot's bible, and moved up to become editor and publisher. He is now a divisional vice-president of Ziff-Davis Publishing Company.